FIGHTING FROM ABOVE

WAYS OF WAR
David J. Ulbrich and Matthew S. Muehlbauer, Series Editors

FIGHTING FROM ABOVE

A Combat History of the US Air Force

BRIAN D. LASLIE

UNIVERSITY OF OKLAHOMA PRESS : NORMAN

Library of Congress Cataloging-in-Publication Data
Names: Laslie, Brian D., author.
Title: Fighting from above : a combat history of the US Air Force / Brian D. Laslie.
Other titles: Combat history of the US Air Force
Description: Norman : University of Oklahoma Press, 2024. | Series: Ways of war; vol 1 |
    Includes bibliographical references and index. | Summary: "Explores whether the US Air
    Force has a unique way of war and identifies three epochs in air force history to determine the
    answer: the realization of airpower's potential, the dominance of strategic air power, and the
    rise of a tactically dominant style of war"—Provided by publisher.
Identifiers: LCCN 2023036250 | ISBN 978-0-8061-9366-3 (hardcover) | ISBN 978-0-8061-9367-0
    (paperback)
Subjects: LCSH: United States. Air Force—History. | Air power—United States—History. |
    United States—History, Military. | Air warfare—History.
Classification: LCC UG633 .L365 2024 | DDC 358.400973—dc23/eng/20231006
LC record available at https://lccn.loc.gov/2023036250

Fighting from Above: A Combat History of the US Air Force is Volume 1 in the Ways of War series.

The paper in this book meets the guidelines for permanence and durability of the Committee on Production Guidelines for Book Longevity of the Council on Library Resources, Inc. ∞

Copyright © 2024 by the University of Oklahoma Press, Norman, Publishing Division of the University. Manufactured in the U.S.A.

All rights reserved. No part of this publication may be reproduced, stored in a retrieval system, or transmitted, in any form or by any means, electronic, mechanical, photocopying, recording, or otherwise—except as permitted under Section 107 or 108 of the United States Copyright Act—without the prior written permission of the University of Oklahoma Press. To request permission to reproduce selections from this book, write to Permissions, University of Oklahoma Press, 2800 Venture Drive, Norman OK 73069, or email rights.oupress@ou.edu.

For the veterans in my family:

Lieutenant Colonel Dallas K. Stephens Sr.
Lieutenant Colonel Dallas K. Stephens Jr.
Captain Heather L. Laslie
Paratrooper William Wayne Laslie Sr.
Sailor Gail Herring
Soldier Joseph Newton

and also to my brothers-in-arms
Dale Nelson and Michael McNerney

Let us bow low before them and salute them very respectfully. Glory to all these volunteers. Glory to all these noble heroes, these noble forerunners. The nation which bore them is a great nation, and I am sure that Remembrance will keep fresh their names and teach their deeds to its children and children's children.
—Capt. Georges Thenault, *The Story of the Lafayette Escadrille*

Contents

List of Abbreviations	xi
Introduction	1

Epoch I. Period of Discovery, 1907–1941

1. World War I and the Birth of American Airpower	27
2. The Interwar Period and the Beginning of Independence	48

Epoch II. Strategic Dominance, 1942–1975

3. World War II: What Hath Man Wrought?	69
4. A Strategic Air Force	100
5. Strategic Aberrations: Korea and Vietnam	114

Epoch III. Tactical Ascendancy, 1975–2019

6. A Useful Way of War? American Airpower after Vietnam	133
7. Desert Storm and the Balkans Campaigns	149
8. Operations Enduring Freedom and Iraqi Freedom	174

Epoch IV. An Unmanned Future? 2019–

9. A Final Frontier or an Unmanned Future?	189
Conclusion	200
Note on Sources and Acknowledgments	205
Notes	207
Bibliography	227
Index	241

ABBREVIATIONS

AAA	antiaircraft artillery
ACC	Air Combat Command
ACM	air combat maneuvering
ACSC	Air Command and Staff College
ACT	air combat tactics
ACTS	Air Corps Tactical School
ADC	USAF Air Defense Command
AEF	American Expeditionary Force
AFA	Air Force Association
AFHRA	Air Force Historical Research Agency
AFHSO	Air Force Historical Support Office (now Division)
AFSPACE	Air Force Space Command
ASTS	Air Service Tactical School
ATC	Air Transport Command
AWC	Air War College
BFM	basic fighter maneuvers
BSA	Bosnian Serb Army
CAF	combat air force
CAOC	Combined Air Operations Center
CAP	combat air patrol
CAS	close air support
CBI	China, Burma, and India
CBO	Combined Bomber Offensive
CENTAF	US Air Forces Central
CENTCOM	United States Central Command
CHECO	Contemporary Historical Evaluation of Combat Operations
CINCNORAD	commander in chief of NORAD
CINCPACFLT	commander in chief Pacific Fleet

CINCSPACE	commander in chief of US Space Command
CJTF-OIR	Combined Joint Task Force–Operation Inherent Resolve
CONAD	Continental Air Defense Command
C_2	command and control
DARPA	Defense Advanced Research Projects Agency
EMT	energy maneuverability theory
ETO	European theater of operations
FEAF	Far East Air Force
GCI	ground-controlled interception
GPS	Global Positioning System
GWAPS	*Gulf War Air Power Survey*
HADPB	high-altitude daylight precision bombing
HOTAS	hands on throttle and stick
IADS	integrated air defense system
ICBM	intercontinental ballistic missile
ISIS	Islamic State
ISR	intelligence, surveillance, and reconnaissance
JCS	Joint Chiefs of Staff
JDAM	joint direct attack munition
LVG	Luft-Verkehrs-Gesellschaft
MACV	Military Assistance Command–Vietnam
MAJCOM	major command
MATS	Military Air Transport Service
MiG	Mikoyan-Gurevich
MOL	Manned Orbiting Laboratory
NFZ	no-fly zones
NORAD	North American Air Defense Command
ODF	Operation Deliberate Force
PGM	precision-guided munition
Pk	probability of kill
PTO	Pacific theater of operations
RP	route package
SAC	Strategic Air Command
SAM	surface-to-air missile
SEAD	suppression of enemy air defenses
SOF	special operations forces
SPACECOM	United States Space Command
TAB	*Tactical Analysis Bulletin*
TAC	Tactical Air Command
UAV	unmanned aerial vehicle

USAAF	United States Army Air Forces
USAFA	United States Air Force Academy
USSBS	*The United States Strategic Bombing Surveys*
USSF	United States Space Force
WAFS	Women's Auxiliary Ferrying Squadron
WASP	Women Airforce Service Pilots
WFTD	Women's Flying Training Detachment

Introduction

This is a history of the United States Air Force (USAF) and its predecessor organizations from their earliest battles to its modern incarnations of airpower (and more recently space power), and the ways in which the air forces of the United States created an independent way of war over time. I specifically choose to say "*a* history" of the USAF and not "*the* history." This book focuses almost exclusively on the conduct and impact of the air domain during times of military action—that is, aerial combat—but also to a lesser extent on the movement of cargo, the refueling of aircraft, and reconnaissance and surveillance. Within these pages, I also examine and prioritize technology, politics, individual and service rivalries, and logistics. In addition, to understand any military arm's way of war, one must look at policy, organization, equipping, and training—particularly in years between wars—to uncover their impact on combat operations. To that end, this is a book that crosses several areas of study: the human and technological elements inside the air force, and the experience that occurred at the meeting of the two.

This book argues that the USAF has had four distinct and observable ways of war that have changed over the course of four distinct epochs. Airpower historian Phillip Meilinger developed "10 propositions regarding airpower," historian and former director of the Air Force History Program Richard Hallion developed "10 attributes of airpower," and strategist Colin S. Gray produced "Airpower theory in 27 Dicta." I have chosen four epochs. The individual epochs of this history I employ are Epoch I: Period of Discovery (1907–1941) and the creation of roles and missions with bombardment theory and practice assuming its role of primus inter pares; leading to Epoch II: Strategic Dominance (1942–1975), in which the ideas of strategic bombardment ruled the USAF until it was unceremoniously proven false during the Vietnam conflict. This led to Epoch III: Tactical Ascendancy (1975–2019), in which fighter and attack aircraft proved dominant; and we entered into a transitional phase of Epoch IV: An Unmanned Future? (2019–).[1]

By breaking down the Air Force way of war into four distinct epochs, I hope to provide rigor to a concept that has traditionally lacked analytical clarity and demonstrate how US officers and officials conceived of the best and proper way to employ airpower. I will also show how changes in the international environment, institutional developments among and within the US military and US government, and new technologies all created distinct ways of war over time for the USAF. Obviously, in a book like this it will be impossible to mention everything or to cover subjects in great detail. Some readers will likely come away from this book with a list of things not mentioned or perhaps take umbrage with the attention paid to such Air Force missions as cargo, transportation, and airlift or refueling and intelligence and reconnaissance. To that end, it is important to keep in mind that in its focus on the Air Force's ways of war, the book's emphasis therefore resides mostly on preparing for combat and engagement in combat.

It is worth defining here the phrase "way of war," or at least defining how it is envisioned in this work. Obviously, there remains the important political consideration of using the armed forces as the extension of politics. Noted historian and strategist Antulio J. Echevarria II noted in *Reconsidering the American Way of War* that interest in the way of war concept "remains strong today largely because of the desire to understand how the United States, the world's sole superpower, might use military force in the future." Thus, this book is a microstudy of one component of American power: airpower. It focuses solely on the Air Force way of war by tracing the history of the use of force by the air arm through time. By borrowing from Thomas Kuhn's *Structure of Scientific Revolutions*, this book aims to demonstrate the paradigm shifts in the air arm and show how the dominant way of doing things changed through distinct epochs. When faced with a crisis, the USAF reconsidered its assumptions and successfully birthed a new way of war.[2]

This book is a constructed history of the United States Air Force and is an addition to the ever-growing literature on American "ways of war." However, in the writing and revising of this book, I decided that any attempt to define a singular way of war for an organization that is more than one hundred years old was, in fact, unwise, especially since that "way" demonstrably changed over time. Sometimes, the USAF way of war is seen most clearly through technological advancement, sometimes in a more human-focused manner, and sometimes in a combination of both. The prevalence of one way of war inevitably gives way to another that better fits the roles, missions, and contexts of the time. The result in all of these epochs is an attempt to achieve victory through airpower, and to allow the USAF to accomplish any and all missions presented to it. In drawing heavily from the existing scholarship and historiography of the USAF, it is my hope that this book can also be used to discover sources, books, digital media, and other avenues of approach for the study of the United States Air Force.[3]

I have not focused on strategy. For that, I suggest readers turn to the esteemed Colin Gray's *Airpower for Strategic Effect*. This book was never intended to unearth new documents or be steeped in primary source documentation. It was always intended to be, in a way, a long historiographical essay. Not every action or engagement in which the USAF participated will be found in these pages. I look at the events I consider most important to the development of the USAF's way of conducting combat. Inevitably, I probably missed some things. Problematic omissions are my own.

Two factors brought this work to fruition. The first, after the release of *The Air Force Way of War*, the University of Oklahoma Press approached me about the feasibility of broadening my previous limited study of one aspect of Air Force history into a more encompassing and inclusive look at how the Air Force—and its predecessor organizations—waged war over the century of its existence. Did there exist a discernible "Air Force way of war" that could be traced over this time frame? The second factor, aided by the first, was my time teaching an introduction to military history course at the United States Air Force Academy and later as an adjunct professor at The Citadel teaching a graduate-level seminar in the history of airpower.

At the Air Force Academy, all freshman cadets are required to take an introduction to military history course. In teaching this class, I began to wonder how the Air Force envisioned its own history and what lessons and events it wanted to instill into its future officers at this early point in their education and training. From a certain point of view, much of what is found herein are the aspects of Air Force history I deemed most important that I wanted to impart to my students. Thus, both the outline and the general direction of this work began to take shape.

Ways of War

The concept of a singular way of war gained notoriety in Russel Weigley's *American Way of War*, a work generations of graduate students and military professionals have read and argued over. Weigley noted in his introduction that "the strategy of annihilation became characteristically the American way in war." Weigley's work also birthed a subfield in military history that considered the concepts of ways of war from various vantage points. Perhaps most notable has been the reassessment of Weigley's work in Antulio J. Echevarria's

Reconsidering the American Way of War. Reading *The German Way of War, The First Way of War,* and *The Skulking Way of War* shaped my own research and writing, particularly of *The Air Force Way of War.* A more recent contribution to the literature, Matthew S. Muehlbauer and David J. Ulbrich's *Ways of War: American Military History from the Colonial Era to the Twenty-First Century,* was the key to bringing this project forward and turning it from a one-year sabbatical-like attempt at a manuscript into a half-decade examination of the USAF. There is the possibility that those steeped in the history of airpower or the history of the United States Air Force will find nothing new here, yet I hope to provide a new lens through which to view just how and why the Air Force conducts war in the air. This work is in a similar vein to Frank Ledwidge's *Aerial Warfare,* Charles Gross's *American Military Aviation,* and John Andreas Olsen's *History of Air Warfare,* with the exception that this work is focused *only* on the United States Air Force (and its predecessors).[4]

With regard to other histories that focus solely on the USAF, I have used and built upon the works of others, both official and nonofficial. This includes the Air Force Historical Foundation's *The Air Force,* edited by Gen. James P. McCarthy and Col. Drue L. DeBerry; Bernard Nalty's two-volume *Winged Shield, Winged Sword: A History of the United States Air Force*; and Robert Frank Futrell's two-volume *Ideas, Concepts, Doctrine: Basic Thinking in the United State Air Force.* It was with some surprise that I discovered that airpower scholarship focusing on the entire history of the USAF was sparser than I had imagined. This, I believe, is due in no small part to ambiguity surrounding when the United States Air Force can be said to have begun.[5]

A Longer View

This work recognizes that *an* American air force existed long before *the* United States Air Force became a separate and independent service. This is a history that begins not with Air Force independence in 1947, but with the first American airmen to engage in meaningful aerial operations prior to the First World War.

Many questions emerged over the course of this study. The main question seemed clear: Is it possible to discern a unique "way of war" for the Air Force? This question led to others: Would it be possible to separate combat operations in war from those during the interwar periods? Does the Air Force's way of war change significantly between major combat operations versus smaller engagements? Was there a significant difference in actions against state or nonstate actors? These questions, ultimately, led to more questions than answers, but along the way a conceptual framework began to take form that helped explain the entirety of the United States Air Force's combat history. This framework became the basis for the sections or "periodizations" of Air Force history found in this book. These are the epochs of American airpower and the defining ways of war, each with its own "thesis."

I decided early on that I wanted to break the history into workable "chunks" of time, yet something lasting longer than the frame of particular chapters. In the history of the United States Air Force, there have been three distinct periods, or dominant epochs, of airpower around which I formed each thesis: the creation and realization of airpower as a military instrument, or as I call it a "period of discovery," running from roughly 1907 to 1941, the dominance of strategic airpower from 1942 to 1975, and the acceptance of a more tactically dominant style of war or a "blended" concept of airpower, 1975–2019. Finally, there is the possibility that even now the Air Force is entering into or, at the very least, moving toward a fourth epoch that could fundamentally alter its current way of war: that being the era of the unmanned aerial vehicles and a second space age. This becomes increasingly clear with the creation of the United States Space Force in 2019.

I decided early on in the research for this study that the only unifying through line from epoch to epoch was the men and women who flew the machines as well as the men and women who maintained them. At the end of the day, it was the willingness of humans to climb into aircraft and fly forward into combat that represented a continuous style of war through the different epochs. This will probably strike some readers as a glib telling of the Air Force's story or a reductionist view of its history. However, the Air Force's way of war might have changed through time with emphasis or preference in any given period being held by a particular airframe, capability, or dominant paradigm, but it was the men and women who flew the machines that truly characterized the Air Force's way of war. It is therefore also a story about these "airmen" (if I may be allowed a nongendered use of an accepted term). But what linked these men, and later women, who climbed into aircraft to conduct aerial warfare? What linked the generations of airmen? What theme or theory unified them?

In 2012, Paula Thornhill wrote a thought-provoking paper about this very issue. Thornhill, a retired USAF brigadier general who now researches and teaches at the Johns Hopkins School for Advanced International Studies, identified five different eras in the history of the Air Force and concluded that it was a desire to go "Over Not Through" that linked airmen. While not disagreeing with her, I find the name of another of her eras more appropriate to characterize the unity of the Air Force over time: "Victory through Air Power" best defines the history of the United States Air Force. First presented in the 1942 book by pilot, writer, and theorist Alexander P. de Seversky, *Victory through Air Power*, the concept represents the driving force and intellectual underpinnings of the USAF throughout its history, and thus also represents a fitting title to this history (though the editors at the press found that title a bit hubristic). Besides, as will be seen, the USAF more commonly believes in victory *with* airpower than victory *through* airpower alone, despite continued (meaningless) debates about whether airpower can win wars on its own. There is a striking difference in which prepositions you use.[6]

Each chapter will look at a particular conflict or periodization and, thus, will seek to answer a similar question: How was the Air Force's (or its predecessor organizations') way of war demonstrated? What did the Air Service, Army Air Forces, or US Air Force hope to accomplish in each war or conflict? What were the results? How did each and every period and conflict link to the one before it? In asking and hopefully answering such questions, I prioritize unifying factors in the history of the USAF.

The study of American airpower begins long before Congress granted the United States Air Force its independence from the US Army in the National Security Act of 1947. There are several dates that could be used for the American air arm's "birthday." On 1 August 1907, the US Army established the aeronautical branch of the Signal Corps. That would seem to be a feasible birthday. An argument also could be made for the opening of the Wright Brothers Flying School at present-day Maxwell Air Force Base, Alabama, in 1910. An even stronger case can be made for the Army's (and now the Air Force's) oldest flying unit, which originated on 5 March 1913. Nevertheless, for the purpose of this book, a working date of 15 September 1914 will be used. For on this date, the first combat unit composed of American fliers organized itself in France. Known later and for all eternity as the Lafayette Escadrille, the American survivors of this unit became the core cadre of later American fighter units that would conduct air operations upon America's entry into the First World War. Other Americans not specifically flying with the Escadrille or Lafayette Flying Corps took up the call to arms as well. Therefore, I maintain that the American Air Force was born not in September 1947, but in September 1914.[7]

The "Stormy Petrel of the Air," as Brig. Gen. William "Billy" Mitchell was known, said in his book *Winged Defense* (1925) that "air power may be defined as the ability to do something in the air." Airpower theorists and historians have debated that intangible "something" for more than a century. Airpower, at times, has been viewed along a continuum from the ultimate form of warfare to an anathema that has brought destruction and death but not decisive or conclusive victory. Even the very term "airpower" has been used in both a positive and a pejorative sense. Airpower has been said by some to have "limits" that govern its execution, and at the same time it has been viewed by others as having a singular ability to decide the outcome of wars that other force components, such as land and sea power, cannot claim.[8]

Mitchell later and more succinctly stated that "air power has brought with it a new doctrine of war which has caused a complete rearrangement of the existing systems of national defense. . . . The future of our nation is indissolubly bound up in the development of air power." One of the most important aspects of the Air Force is its tendency to justify itself by things expected in the future rather than by things already accomplished in the past. Here, for example, Mitchell claimed for the air force the sort of effectiveness that it simply did not have at the time he wrote the words.[9]

It must also be added that the concepts, theories, and doctrines of airpower have often been viewed through the lens of technology. The advancements in the military application of airpower are interwoven with technological advancement. The two run parallel courses. However, for both members of the US Air Force as well as historians, the technological aspect often overshadows other equally important contributions of developments in tactics and training. The focus on "things" over "theory" often hindered Air Force intellectual development and stymied the evolution of existing theories. What follows in this introduction is a general overview of my conceived epochs—a basic "sweep," to use a tactical term, through the material that will be explored in further detail in later chapters.

Epoch I: Period of Discovery (1907–1941)

The four-volume compilation *The U.S. Air Service in World War I*, by Air Force historian Mauer Mauer, overflows with recollections of the officers of the American pursuit and bombardment squadrons. While Aviation Section leaders, notably US Army general Benjamin Foulois, focused on organizational conflicts, the fliers focused on tactics and doctrines. One such pilot was Capt. James Meissner, who stated of his own preparation for combat in the First World War that "the training schools have not sent men to the front prepared for their work. . . . The combat instruction given to them is based on the wrong idea. They have been led to believe that their combat principles involved individual combat principally whereas individual combat is a very rare occurrence."

Clearly, the focus on the need for realistic training is as old as airpower itself, and Meissner showed great concern that the pilots arriving at aerodromes were ill-prepared to employ their technologically advanced aircraft. Still, it is instructive to note that pilots have always recognized shortcomings in their training. Acknowledging this allows us to appreciate the fact that those who participated in the "last" war often set the standards and training regimens for the next war.[10]

The official history of the British army found bombardment to have been "without important results" during World War I. It should be readily admitted that despite public focus on airplanes and the airmen who flew them, their contributions to the outcome of that war itself were, at best, negligible. The most important aspect of the air war in Europe was the defining of roles and missions: reconnaissance, bombardment, and pursuit (later, fighter). As historian Lee Kennett stated in *The First Air War*, combat aircraft in the First World War developed "according to the needs of battle more than according to a doctrine or in some deliberately chosen direction." The "needs of battle" began with aerial reconnaissance. Early aircraft focused on roles of artillery spotting, observation, and reconnaissance, and it was from these roles that aerial combat grew. With few exceptions, by the end of the First World War, deep divisions and distinctions were drawn as to the utility of airpower's components. The pursuit pilot might have held the glamour, but the future of aerial warfare was vested in the efficacy of strategic aerial bombardment. The schism between fighter pilot and bomber pilot has never entirely healed.[11]

If the study of American aerial combat begins with the Lafayette Escadrille and other Americans flying for foreign units, then the study of the development of American airpower doctrine begins—in some ways this is regrettable—with William Lendrum "Billy" Mitchell. Mitchell was no theorist; "propagandist" is a better term. He had an ability to rub his colleagues and superiors the wrong way and made enemies out of many of them. Seeing the success of the British Royal Air Force (RAF) at gaining its independence, he wanted the same thing for an American air force. This became his ultimate goal no matter whom he offended along the way. Still, doctrinally Mitchell recognized that "air forces will strike immediately at the enemy's manufacturing and food centers, railways, bridges, canals and harbors" and that since airpower was so radically different than what came before, an air force should be led by "air-minded individuals in their own service." After his court-martial in 1925, Mitchell became a martyr for Air Corps members in the 1920s, 1930s, and 1940s, and his admirers and defenders all "hung their hats" on Mitchell's belief in an independent air force. This was clearly demonstrated during doctrinal development in the interwar years.[12]

The intellectual development of American airpower took place at the Air Corps Tactical School (ACTS), located at Langley Field in Virginia, from 1920 to 1931, and thereafter at Maxwell Field in Alabama. Historian Alan Stephens called the Air Corps Tactical School "a vibrant, innovative environment, in which the evolving and often competing schools of air power doctrine . . . were argued with a passion." The list of those who both attended and later taught at the school and became important figures during the Second World War includes Claire Chennault, George Kenney, Joseph McNarney, Muir S. Fairchild,

Laurence Kuter, and Hoyt Vandenberg. Yet, if the school was vibrant and innovative, it was not always a place where differing opinions were accepted. Stephens is right in his assertion that differing schools of thought were argued and discussed seriously at the ACTS; however, one was primus inter pares: bombardment. So deeply ingrained and accepted was bombardment that tactical aviators, most notably Claire Chennault, found themselves maligned, sidelined, and left out.[13]

The main concept and primary tenet that emerged from the ACTS during the 1920s and 1930s, and endorsed by its staff, was that airpower should be viewed as an inherently offensive weapon; even a defensive use of airpower, as in a patrol over one's own airfield, becomes offensive once enemy aircraft are engaged, as opposed to how ground forces could operate in a purely defensive manner. The second tenet, first quoted by British Prime Minister Stanley Baldwin in 1932 before Parliament, became a mantra: "The bomber will always get through." The importance of the development of bombardment aviation cannot be overstated. Pursuit aviation was justified only as a means to provide support for bombers, and it was subsumed by bombardment theory. The bombing tests by Billy Mitchell in July 1921 off the Virginia Capes proved to many that the future of airpower was vested in the advancement of bombardment theory despite the myriad problems with the conduct of the tests themselves.[14]

If there was one theory that became accepted doctrine and found great tactical use in the curriculum of the ACTS, it was high-altitude daylight precision bombing of the enemy's industrial base. Although the intentional bombing of civilian targets was never officially endorsed as a means to end conflicts by the United States Army Air Forces (USAAF), during the Second World War it was viewed as a desirable secondary outcome of the bombing of industrial targets, and it was supported prior to the war by original airpower theorist Giulio Douhet. Douhet believed, perhaps paradoxically, that bombing of civilians would lead to a decrease in overall casualties because, as he put it in 1921, "the time would soon come when, to put an end to horror and suffering, the people themselves, driven by the instinct of self-preservation, would rise up and demand an end to the war." Surveys done after the war clearly proved that this had not been the case. However, airpower advocates at this time latched on to bombardment as the future of air warfare. This belief in strategic bombardment during and after the war was so deeply rooted, that it took the Air Force many decades to shake itself loose from this concept.[15]

By 1926, ACTS field manuals were giving preeminence to bombardment as a means to destroy enemy aircraft on the ground or in their factories before they could be engaged in the air. With this focus, the air arm began to experiment with the doctrines that would give rise to the second epoch. The next logical step for the proponents of this view of the primacy of bombardment was that, if bombardment could destroy the enemy's air forces and *resources* on the ground, then perhaps that method of attack could also destroy the enemy's *will* to fight. This idea was pervasive with many of the ACTS's instructors and students. As bombardment adherents preached and the importance of bombing theory rose, those in favor of pursuit and attack theories diminished. One notable exception was Capt. Claire Chennault, who advocated for reliance on pursuit aviation to provide protection for the bombers. As historian Martha Byrd described this hardening of opinion regarding the superiority of bombing over pursuit in *Chennault: Giving Wings to the Tiger*: "Pursuit seemed to be of steadily diminishing importance."[16]

Chennault approached debates with bombardment advocates with "the fervor of an evangelist as he sought to save bombardment from itself." Even the tactically minded Pete Quesada did not fall in with Chennault. As historian Thomas A. Hughes stated in *Over Lord: General Pete Quesada and the Triumph of Tactical Air Power in World War II*, "Quesada did not buy into Chennault's notions and found the clean thought and internal logic of bomber ideas more sound." The invention of radar later validated Chennault's insight that pursuit aircraft would indeed be able to interdict bomber aircraft. By that time, however, Chennault's disputes with his colleagues had become so intense that he resigned his commission in April 1937. Two months later, Chennault was in China, at the behest of the Chinese government, a move he had pondered for some time and that was sped along by his feelings of marginalization by the bombardment proponents at ACTS. In China, Chennault led the famed Flying Tigers in the air war against Japan. Did the study of pursuit aviation in America die when Chennault left ACTS? No. His post as chief of the Pursuit Section was taken over by Capt. James E. Parker and later Capt. Hoyt S. Vandenberg. Still, bombardment reigned supreme. And so it was that the use of airpower during the First World War—the defining of roles and missions, and the creation of an "air force" that could "do something in the air"—constituted the way of war for the US Army's air arm. Finally, the development of bombardment theory in the interwar years and its conduct in the Second World War saw the transition to the second epoch in American airpower.[17]

Epoch II: Strategic Dominance (1942–1975)

Because of their wisdom demonstrated over the course of many years, such classical military theorists as Sun Tzu, Clausewitz, Jomini, Mahan, and Corbett continue to be studied in military schools. The study of airpower, however, is comparatively new. Over the last one hundred years, proponents of airpower theory have struggled with the need to be found relevant in their time. The names Douhet and Mitchell are known not only for their links to airpower theory in general, but also for their advancement of a particular airpower agenda. Airpower theorist and zealot Alexander de Seversky claimed that victory in World War II was possible *only* through the use of long-range strategic airpower. In his book *Victory through Air Power* (1942), de Seversky warned of dire consequences for *not* taking immediate action. He wrote, "A realistic understanding of the new weapon, of its implications in terms of national security, of its challenge to America, is not a matter of choice. It is the very condition of national survival." De Seversky often invited conflict.[18]

De Seversky is pertinent to the discussion of how deeply wedded the American Air Force and American public became to bombardment primarily because his writing came to the fore during the Second World War, bringing the idea of airpower into the homes of ordinary Americans. The greatest criticism lodged against those who opposed de Seversky's often outrageous claims was their inability to prove him false. When Alfred Thayer Mahan wrote *The Influence of Sea Power upon History* in 1890, the importance of naval strength was accepted fact, with thousands of years of evidentiary support behind it. Conversely, when de Seversky wrote *Victory through Air Power*, people had been flying in heavier-than-air aircraft for fewer than forty years, and the airplane had only been used in combat for a fraction of that time. Newspaper columnist David Brown wrote in his two-part critique "Victory through Hot Air Power" that "the defense of *Victory through Air Power* . . . rests upon the notion that total air power has never been tried. It was thus impossible to refute Seversky's claims of infallibility."[19]

The inability to refute de Seversky's perspective was a common theme that followed the publication of *Victory through Air Power* and the dominant strategic bombardment theory for decades. In the second edition of *Makers of Modern Strategy*, published in 1969 (originally published in 1941, prior to de Seversky's work), one critic said, "Much of Seversky's book is frankly prophesy [*sic*] but even when he writes in the future tense he leaves himself open to the charge of being unduly optimistic about the rapidity with which his visions of future aircraft and their performance will be realized." De Seversky was another Billy Mitchell looking to the future of what an independent air force could accomplish.[20]

There were clearly two viewpoints on *Victory through Air Power*, and these developed into two opposing camps, simply: those who supported de Seversky's ideas—many of whom previously supported or were acolytes of Billy Mitchell—and those who opposed them. Historian Charles Beard, a stated isolationist and adversary of America's involvement in the Second World War, found de Seversky's ideas pertinent: "In my opinion this book is more important to America than all the other war books put together."

The subsequent film version of the book made by Walt Disney, along with the book itself, helped link aspects of airpower theory, such as strategic bombardment, to public opinion, and the debates for or against airpower and its utility played out not only inside the military, but also around the dinner tables of America. The phrase "victory through air power" became locked into the lexicon of the United States Air Force, even if few remember today the book or the film.[21]

For Air Service leaders, strategic bombing theory reigned supreme during World War II, and the US Army Air Forces and the UK's independent RAF of Great Britain dove into attacking Germany's—and later Japan's—industrial centers. The new long-range bombers bristling with guns punching out of every hole were perceived by the proponents of bombardment as equal to or better than any single-seat fighter they would encounter. The prevailing thought across the European continent, as well as in America, was that larger multiengine bombers maintained a technological and qualitative edge—they flew faster and higher—and were therefore incapable of being intercepted or shot from the sky. In the United States Kenneth Walker, instructing at the Air Corps Tactical School prior to America's entry into the war, noted, "Military airmen of all nations agree that a determined air attack, once launched, is most difficult, if not impossible to stop."[22]

Larry Kuter, another ACTS instructor, stated, "We realize that we, in the Air Force section, may be extravagant air enthusiasts. It has been more than once inferred that the exponent of the effectiveness of air power has the ability to magnify his own powers and minimize the opponent's powers approaching dangerously near the brink of mania. We have made an honest effort to temper our zeal." However, the bottom line was that the dominant thinking of the time held that there was no force on earth capable of stopping an aerial attack once it began. As Kuter put it, "Shells from all guns miss small targets quite frequently. A maneuvering airplane at long range or high altitude is the most difficult target ever assigned to any piece of artillery"; there was no way to intercept incoming aircraft short of having patrols flying constantly at all hours of the day. Of course, the development of radar and other air defense mechanisms would fundamentally alter this calculus.[23]

In practice, these underlying assumptions about long-range bombers proved to be wrong. A bomber was not equal to a single-seat fighter in aerial combat, and advances in antiaircraft fire proved that bombers would not always get through the enemy lines. More importantly, the efficacy of strategic bombing theory was never what its proponents predicted. As historian of technology Barton Hacker put it in *American Military Technology*, "In contrast to the theorists of armored warfare, the prophets of strategic bombing misread the future in almost every respect. Even apart from its moral implications, air attack on enemy industry and population proved far harder to mount, more costly to maintain, and less productive of results than they had supposed." Nevertheless, it represented the dominant thought at the time.[24]

In the history of the USAAF, tactical aviation played an important role in each theater during World War II but has traditionally been overshadowed by the strategic bombers. Air Force leaders focused on strategic bombing to cause political upheaval in Germany and Japan. The bombing campaigns primarily attacked industry but, to a lesser extent but every bit as real, also attacked the populace. Writing in his diary about the pilots who flew the bombing missions against Japan, Hap Arnold stated, "There is no feeling of sparing any Japs here—men, women, children—gas, fire, anything to exterminate the entire race exemplifies the feeling." That being said, the Allies' tactical airpower—fighters, attack, small and medium bombers, and dual-role fighter-bombers—played an important role across the globe by aiding in the destruction of Axis tactical airpower assets and directly supporting the ground forces.[25]

In the European theater of operations (ETO), Gen. Pete Quesada's tactical forces flew fewer sorties and dropped considerably less iron than their bomber brethren, but it was the tactical forces that kept the German forces so disorganized that, at times, they could barely operate. Despite strategic raids against petroleum, oil, lubricant, and aircraft production facilities, it was not strategic bombing that dismantled the Luftwaffe, but rather prolonged engagements with tactical fighters, including the P-47s and P-51s. *The United States Strategic Bombing Surveys* (*USSBS*) compiled after the end of the war clearly indicate that once tactical airpower was given a freer hand to engage the Luftwaffe while escorting Allied bombers, the number of enemy aircraft destroyed increased dramatically. However, as historian Thomas Hughes stated, "History quickly and easily forgot the contributions of Pete Quesada and others associated with tactical air power in World War II." Though the successes of tactical airpower began to grow, it was bombardment that remained king. An apocryphal quote attributed to Curtis LeMay summed it up, "Flying fighters is fun. Flying bombers is important."[26]

Strategic bombardment played an even more important role in the Pacific theater of operations (PTO), even if the total tonnage of bombs dropped was considerably less. The fall of 1943 saw the first bombardment of Japanese industrial targets in Manchuria and Kyushu with the new technological marvel the B-29, flying from bases in China. At this time, the islands of Guam, Saipan, and Tinian were not under Allied control, and flying from China was the only way to reach the Japanese "inner zone industries." Ongoing bombardment contributed to the perception among many Japanese civilians that the war was lost even before the atomic bombings of 1945.[27]

It is important to recognize that the *USSBS* clearly stated that "certainly prior to 31 December 1945, and in all probability 1 November 1945, Japan would have surrendered even if the atomic bombs had not been dropped, even if Russia had not entered the war, and even if no invasion had been planned or completed." Obviously, this was written in retrospect and no Air Force commander, down to Col. Paul Tibbets, who captained the *Enola Gay*, ever intoned that dropping the bomb was not the correct course of action given the information available at the time. However, to the American people and Air Force leaders, the atomic bombings became the primary cause of Japan's capitulation.[28]

It is important here to mention in some detail the achievements of training, tactical, and attack aviation that contributed to success in the PTO. Tactical aviation in the Pacific, at least in popular histories, is almost entirely neglected in favor of strategic bombardment, and the treatment of tactical aviation in academic works is limited at best. However, this pertains only to the USAAF and cannot be said of the US Navy's carrier aviation assets that broke the back of the Imperial Japanese Fleet. Nearly lost to all but a handful of military historians are the important contributions made to tactical aviation by Gen. Douglas MacArthur's lead airman, Gen. George Kenney. Thomas E. Griffith's work *MacArthur's Airman: General George C. Kenney and the War in the Southwest Pacific* showed him to be an exceptional air commander who used both strategic and tactical aircraft to achieve operational-level successes in his corner of the Pacific theater. Kenney's unconventional methods including use of bombers for low-level attack and his advocacy of "skip bombing" paid dividends in battle, for example, at the Bismarck Sea. Kenney was, in a certain sense, a prototypical air component commander the likes of which would not be seen again until Desert Storm.

Kenney himself, even if a bit self-aggrandizing in his autobiography, recognized the importance of realistic training scenarios for his airmen. As Griffith stated: Kenney's motives for additional training were partly humanitarian; there was also a practical side to the measures. Better training increased the morale of the aviators and led to better combat results. In addition, Kenney's emphasis on training translated into fewer aircraft losses on combat missions, thus preserving planes for future operations. In short, training was a prudent investment for the future.[29]

However, Kenney's contributions in one corner of one theater during World War II could hardly hope to compete with the awesome power of the atomic bomb and the powerful hold it would gain in the minds of the American people and their military leaders. Even Kenney admitted, "There was no question about the destructive capabilities of this new and terrifying weapon."[30]

In November 1944, the Air Force commissioned the *USSBS* to study the outcomes and effectiveness of the World War II bombing campaigns. The

USSBS served as the Air Force's validating documents after the war ended. The results of the bombing campaigns in the European theater were clear: "Allied air power was decisive in the war in Western Europe. Hindsight inevitably suggests that it might have been employed differently or better in some respects. Nevertheless, it was decisive." As far as strategic bombing in the PTO was concerned, the bombing survey did state that, in all likelihood, Japan would have surrendered in the fall of 1945 even if the atomic bomb had not been used. It is important, however, to point out that the *USSBS* served the interests of the Army Air Forces.[31]

To say that the atomic bomb changed everything in warfare is a gross overstatement. That its creation and use affected the development of future air forces would be an understatement. While the advent of nuclear weapons may have hastened the end of the war, the changes the new weapon had on doctrinal developments in the post–World War II Air Force had, in turn, an impact on the development of aircraft as well. If anything, nuclear weapons kept strategic bombardment theory at the forefront of airpower theory and doctrinal development. It was the atomic bombings of Hiroshima and Nagasaki that cemented the primacy of strategic bombardment in the minds of the American populace. As Michael Sherry stated in *The Rise of American Air Power* (1987), "Almost as soon as Hiroshima was destroyed, the reflex reaction of most observers was to regard the atomic age as revolutionary and the previous history of air war as irrelevant, just as earlier commentators on the bomber tended to dismiss the previous history of warfare." As tactical airpower historian Thomas Hughes put it, "The newly independent US Air Force stressed strategic bombardment at the expense of tactical accomplishments in an effort to carve a role for air power in the nuclear age." Surely air forces equipped with nuclear bombs could inflict the type of desired damage on the industrial base of an enemy that would lead to direct capitulation. Delivery of nuclear weapons would remain the focus of strategic and tactical forces for the thirty years following World War II; even fighter aircraft would need to prove the ability to deliver nuclear payloads. With Air Force bombers being the only vehicles capable of nuclear delivery for years to come, Strategic Air Command was essentially the only game in town, and Tactical Air Command soon mirrored the dominant command. The United States Air Force was initially created and existed solely to deliver atomic weapons against the enemies of the United States.[32]

After the Second World War, a complicated international scene with many major players was gradually replaced by the bipolarity of the Cold War, and to meet this new challenge, on 21 March 1946, the United States Army Air Forces created the organization intended to prepare for, deter, and if necessary deliver an atomic attack against the Soviet Union: the Strategic Air Command (SAC).

The creation of the United States Air Force in 1947 with the Strategic Air Command as its singularly dominant "major command" overshadowed developments in the Tactical Air Command.

The Air Force's answer to the growing Soviet threat was to expand the force that was capable of long-range strategic bombardment. The perceived validity of strategic bombing during World War II held by US Air Force leaders, coupled with the decisiveness (and short-term American monopoly) of nuclear weapons and the increasing tensions of the Cold War, allowed SAC, at least after Gen. Curtis LeMay took command in 1948, to grow nearly unhindered by budgetary constraints. The B-29 quickly gave way to physically larger and, with the advent and inclusion of jet engines, technologically superior bombers, including the B-36 Peacemaker, B-47 Stratojet, and B-52 Stratofortress. SAC's raison d'être, as detailed in an early mission brief, was to conduct "long range offensive operations in any part of the world independently . . . and provide combat units capable of intense and sustained combat operations employing the latest and most advanced weapons."[33]

Strategic Air Command's future looked promising as President Dwight Eisenhower's administration took office. The new administration's "New Look" strategy was codified in National Security Council Document 162-2, which stated that the nation needed "a strong military posture, with emphasis on the capability of inflicting massive retaliatory damage by offensive striking power." This doctrine played directly into SAC's strength—namely, its singular ability to deliver nuclear weapons through the use of the strategic bombing force. From the beginning of General LeMay's tenure as commander in chief of Strategic Air Command, or CINCSAC, the branch's strength dramatically increased in both the number of wings and groups as well as the number of aircraft allotted to each.[34]

Leaders of Tactical Air Command (TAC), also established in 1946, did not help their own organization by focusing on the atomic mission as well instead of staying focused on the application of tactical airpower. TAC leaders worked with senior Air Force leaders in Washington, most of them former bomber pilots, to ensure they got at least some share of New Look funding in the 1950s. As Conrad Crane cleverly stated in *American Airpower Strategy in Korea, 1950–1953*, "[Gen. Otto Paul] Weyland and his TAC successors struck a Faustian bargain with the atomic Mephistopheles, transforming the organization into a 'junior SAC' concentrating on the delivery of small nuclear weapons." As such, the Air Force developed, over the course of the 1950s, an array of fighter aircraft, the Century Series, that were meant to perform less as fighters and more as specialized vehicles for delivering tactical nuclear weapons. For their part, the oft forgotten North American Air Defense Command (NORAD) and

USAF Air Defense Command (ADC) existed to protect SAC bombers and act as trip wires against Soviet attack to ensure American bombers had enough time to take off. The ADC also held the mission to intercept incoming Soviet bombers.[35]

Korea

The opening of the Korean conflict in 1950 found the Air Force ill-prepared for the task at hand, including close air support of Army units on the ground and air-to-air operations against the new Mikoyan-Gurevich (MiG)-15. Although the USAF had an initial advantage over the North Korean Air Force, this changed once Stalin began flowing newer equipment to his Communist ally. American airpower in Korea became outmatched both numerically and technologically. Straight-winged American tactical fighters, such as the F-80, were no match for the sleeker, swept-wing MiGs. Historian Conrad Crane noted that, once introduced, the MiG-15 "outclassed all UN [United Nations] aircraft then in Korea, flying more than 100 miles an hour faster than the F-80."[36]

In response to the Korean crisis, the US Air Force decided to deploy a sizable contingent not to Japan or South Korea, but to the European theater to ensure that the Soviet Union did not take advantage of any perceived US lack of commitment in that region. The USAF remained convinced that Korea was a sideshow. Although the Air Force buttressed its assets in Korea, including fighters and bombers to the Far East Air Force (FEAF), senior Air Force leaders, notably Curtis LeMay, viewed the Korean conflict at first as an anomaly, and American leaders saw Europe as the most likely place for the next Communist incursion; they did not want to be caught off guard when the Soviets moved. LeMay noted to an NBC reporter that "SAC was the USA Sunday punch and that every effort must be made to make sure that it stayed intact and able to strike and not be pissed away in the Korean War." The Soviet Union had no intention of actually invading Western Europe, and Air Force leaders not only missed the importance of the air war over Korea but myopically focused on the wrong theater.[37]

The Korean Peninsula fell under the aerial jurisdiction of the FEAF commanded by Gen. George Stratemeyer. The bulk of the aircraft in the command were B-29s and F-84s, which were quickly outmatched in November 1950 with the appearance of the Soviet-made MiG-15 in the skies over Korea. Small and nimble, the MiGs set up effective combat air patrols (CAPs) over the bridges connecting China to North Korea. These roaming patrols of MiG-15s, sometimes numbering in the hundreds, provided an effective defense of the North Korean–Chinese border. Engaging the MiG-15s was a difficult proposition for both fighter and bomber crews. With their home bases located inside China and therefore beyond the aerial jurisdiction of American pilots, the MiGs were able to take off and gain altitude on their side of the border before sweeping down on bomber formations and using the gained energy in their dive to either regain altitude (known as a yo-yo maneuver) or use the speed gained in the dive to turn and cross back into Chinese airspace. In response to the MiG-15, the F-86 Sabre was pushed into service on the Korean Peninsula ahead of schedule. Although the MiG-15 outclassed the new Sabre jets in almost every way, including speed, altitude, and turning ability, the F-86 did have superior armament and flight controls. To counter the F-86, the North Koreans and Chinese began flying their most experienced pilots and continually changed their tactics. The area close to the Chinese border became known as "MiG Alley."[38]

These new enemy tactics included flying in extremely large formations of fifty or more MiGs, which the American airmen called "MiG trains." Since they were engaging such a large number of enemy aircraft, the American pilots could not control the battle, nor could they bar all MiGs from pressing on to confront American bomber formations. The MiGs also created the tactic of "boxing in" American airplanes as they attempted to exit an area once their fuel ran low. Despite these new tactics, by the end of the war American airmen had become quite adroit at destroying enemy MiGs. Overall, the Americans lost more aircraft than did the Chinese and North Koreans, although many of those losses were due to ground fire or mechanical failure. American airmen continued to quote a 7-to-1 kill ratio in direct air-to-air combat, but if one lesson should have been learned from Korea, it was that decisive gaining and maintaining of air superiority over the country—something the Air Force never did—remained the most important thing to do in any military campaign. The lessons learned from the war in Korea from an aviator's perspective were never codified into doctrine, and the Air Force as a whole viewed the aerial combat on the Korean Peninsula as anomalous in kind, unlikely to recur. As Crane stated, "The Air Force, and TAC with it, soon returned to its focus on general nuclear war," and thus, the dominant epoch of bombardment continued despite indications that the capabilities of tactical airpower were clearly rising.[39]

Air Force senior leaders believed the conflict in Korea to be a military anomaly and continued to focus on the Soviet threat. The counter to this threat was the ability of US forces to deliver nuclear weapons. Thus, the USAF's development of small fighter aircraft in the 1950s and 1960s concentrated on the ability of these aircraft to penetrate air defense systems and deliver their payloads. Little to no consideration was given to gaining air superiority, aerial combat against other aircraft, suppression of enemy air defenses (which continued to be a growing threat), or close air support to ground forces. With the air war in Korea effectively ignored in doctrine and training, the US Air Force entered Vietnam woefully unprepared to meet and contest, much less defeat, the aerial threat it faced from small, maneuverable MiG fighters and from surface-to-air missiles.[40]

Lee Kennett, though discussing World War I in his book *The First Air War* (1991), made a statement that applies rather nicely to airpower in general: "One could make the argument that military aeronautics suffered almost as much from the attentions of its partisans as from the attacks of its enemies. This is because its partisans rather consistently oversold it." Every would-be available lesson from World Wars I and II and from Korea indicated to the US Air Force leaders that the future of aerial warfare was indissolubly bound to strategic bombardment. Air Force leaders were also bound to strategic bombers, though by a more nostalgic thread. As historian Kenneth Werrell said in 1976 in his article "The Air Force and the Future of the Strategic Bomber," the "B-29s, B-50s, and B-36s... B-47s and B-52s.... were the aircraft that the top Air Force leadership flew and with which it identifies.... All the Air Force's Chiefs of Staff have flown bombers."[41]

Importantly, throughout the Vietnam War, these same men became chiefs of staff *because* they had flown bombers. Strategic bombardment theory was bound to Strategic Air Command's ability to deliver nuclear weapons. Tactical Air Command's need for money and a mission was bound to SAC. Therefore, TAC was also bound to strategic bombardment theory.

Vietnam

American involvement in Vietnam changed all of this. Vietnam proved to be the high-water mark for strategic bombardment and the final days of Strategic Air Command's dominance in the United States Air Force.

The Air Force's second epoch, the age of bombardment, traces its lineage directly from Italian theorist Douhet to Mitchell to the "father of the Air Force" Henry "Hap" Arnold to World War II bomber and later SAC general Curtis LeMay. In each of these persons, there was a reliance on and belief in the efficacy of strategic bombardment as the best means to achieve victory. The leaders of the Air Force relied on bombardment, and the American people were overwhelmingly presented this concept in such films as *Victory through Air Power* (1943).

Still, occasionally an innovative thinker or even group of thinkers comes along who can "overhaul the institutions meaningfully." For that to happen, though, the thinkers must find and locate the problems inherent in the institution and then set about systematically to correct the problems, often to the outrage of those holding to the dominant paradigm. The problems in the US Air Force—mainly the overreliance on strategic bombardment doctrine—were found and confronted by the generation of officers who experienced combat in Vietnam. The epoch of strategic bombardment waned and a third epoch, the age of dominant tactical airpower, arose in the skies over the jungles of Vietnam.[42]

Epoch III: Tactical Ascendancy (1975–2019)

By 1965, the Air Force had not managed to gain enough confidence in its own institutional existence to work in a meaningfully cooperative way with the other armed services. The Air Force failed to entertain any ideas about how to operate in a manner outside its preconceived notions of how airpower worked. Air Force leadership dismissed any significant reforms to tactical operations out of hand, because to reckon with those particular ideas would have felt like signing the death warrant for the service's independence that they had won on the backs of strategic bombers. However, during Vietnam, it became increasingly obvious that the dominant way of war for the USAF through the use strategic bombardment had massive flaws, and the fliers of fighter aircraft operating day in, day out over South and North Vietnam, Laos, and Cambodia began to subvert the dominant paradigm.

The Vietnam conflict proved to be a watershed event for the United States Air Force as it was for the nation as a whole. It also proved to be the culmination of the doctrine of strategic bombardment. The USAF entered Vietnam with an overreliance in the belief that strategic bombardment could win any conflict where it was applied. Because of this doctrinal overreliance on strategic bombardment, the development and training of the tactical air force had suffered neglect. TAC spent the 1950s turning itself into a junior SAC, focused on nuclear delivery or intercepting Soviet bombers. The command's tactical prowess atrophied. Despite later assertions of its effectiveness, the December 1972 Linebacker II operation flown over Vietnam conclusively proved the futility of flying heavy bombers into the jaws of an integrated air defense system. The organization of airpower during the Vietnam conflict also kept it from providing a unified effect. The air war in Vietnam comprised an in-country air war in South Vietnam, a strategic bombing campaign in North Vietnam, an air-to-air war in North Vietnam, and special operations incursions into Laos and Cambodia. Overlaying these multiple air wars was a division of the country into seven separate route packs, which further hindered airpower's effectiveness. Rather than allow US Navy and Air Force assets to work together, this approach retarded their effectiveness and led to competition rather than cooperation. The two services battled each other as much as they battled the enemy. William Momyer, the 7th Air Force commander under Gen. William

Westmoreland, noted in his book *Airpower in Three Wars* that he was never successful in gaining command and control over the disparate forces in the region. Historian Mark Clodfelter viscerally dissected the role airpower played in 1972 in his book *The Limits of Airpower*. Finally, despite the USAF having a pronounced asymmetric advantage in technology during the conflict, American airmen suffered under a lack of preparation and training prior to engaging in combat operations. The Air Force entered Vietnam flying advanced fighters, bombers, and support aircraft, but these advantages did not provide results due to the failings in doctrine, organization, and training. Personnel, equipment, and technology were not effectively welded together.

The Air Force set about fixing the failings of doctrine after the Vietnam War ended. In 1975, Tactical Air Command set up the Red Flag exercise to address the training and doctrinal deficiencies encountered in Vietnam. This exercise fundamentally altered how the Air Force prepared for combat. Organizationally, the Air Force did not change much in what we might now call another "interwar period." Although SAC and TAC continued to operate independently, the latter took a more pronounced lead. Despite the lack of self-directed organizational change within the Air Force, forces outside the USAF, primarily the US Congress and the Department of Defense, provided for more effective organizational changes. The Goldwater-Nichols Act forced the services to work together in future operations. This actually played into TAC's hands, as it was already actively cooperating with the Army's Training and Doctrine Command. Inside the USAF, fighter pilots began to assume leadership positions, culminating in Gen. Charles A. Gabriel becoming the first pure fighter pilot to become Air Force chief of staff.[43] Finally, in the all-important planning realm, the USAF also received significant doctrinal help from great airpower thinkers such as John Boyd and John Warden.

Thus, when the Air Force again entered into combat in 1991, it was well prepared doctrinally and significantly better trained than in the Vietnam era. Tactical warfare now dominated doctrine. Technologically, the USAF also advanced in this interwar period. New aircraft including the F-15, F-16, A-10, AWACS (airborne warning and control system), and JSTARS (joint surveillance target attack radar system) all moved the Air Force into the fourth generation of air combat. These aircraft were technologically advanced, but the men and women who flew them were also significantly better trained than their predecessors. The training aspect helped shape the battlefields of Operation Desert Storm. It is important to note that the Iraqi Air Force also possessed advanced fourth-generation aircraft, but the Iraqis completely lacked the ability to use them effectively. Desert Storm proved that effective organization, doctrine, technology, and training created a veritable wall of airpower that was difficult if not impossible to scale. American airmen knew better than any generation before them how to use the technology they were given—how to wield it, and how to meld with it.

The 1990s saw continued use of relatively effective airpower during Operation Deny Flight over Bosnia and Herzegovina, Deliberate Force organized against the Republika Srpska, and finally Operation Allied Force, the NATO operation against Yugoslavia during the Kosovo War. Organizationally, the Air Force inactivated the Tactical and Strategic Air Commands on 1 June 1992 in favor of a united Air Combat Command (ACC) that moved all combat aircraft under a single command to provide air options for any conflict. Despite the fall of the Soviet Union in 1991, the USAF continued to prepare for combat operations, and the Balkan campaigns proved the continued need for effective airpower. It is thus at the end of the 1990s that this book will move into the murkiest and far more difficult of eras for the historian to write about, the recent past.

Epoch IV: An Unmanned Future? (2019–)

The final part of this book treads into unfamiliar territory for this author. It includes current events and covers the rise of unmanned aerial vehicles (UAVs), often erroneously called drones, and of space operations. The massive changes in technology in the area of UAVs have brought forth a new weapon in war, but whether or not UAVs are truly revolutionary is yet to be decided. While these weapons have added something new to the arsenal, have they fundamentally altered any aspect of war making for the US Air Force? At the least, they have been very successful in raising questions of morality, legality, and necessity in warfare.

For its part, the Air Force continues to have its cake and eat it too. As UAVs continue to be tasked with missions in permissive environments, the USAF moves forward into the future fielding systems much the same as, but much more advanced than, previous aircraft generations. The F-35, B-21, and KC-46 all signal that the Air Force believes its reason for existence continues to be to gain and maintain air superiority, and thus ensure victory through airpower.

The final chapter of this book explores the role of the United States Air Force in space operations and now the United States Space Force. Long a part of the USAF, space—that is, those in the Air Force who work on space-related issues—has seen its importance wax and wane since the creation of the Air Force Space Command in the early 1980s. Initially, the command included the operations of America's land-based ICBMs, but that mission eventually was shifted to the ACC and finally to Global Strike Command, the reborn Strategic

Air Command. This chapter lands where it does, toward the end of the book and in the last epoch of airpower, precisely because the Air Force has never been entirely sure, or proven conclusively, just where space fits into the Air Force writ large nor what or how space contributes to combat operations. As late as 2017, the USAF fought the congressional effort to create a separate Space Corps within the Department of the Air Force and lacked a clear vision as to whether space operations were indeed combat operations. Any hesitancy on behalf of the USAF was overcome in 2019 with the establishment of the United States Space Command followed by the initial stand-up of a "separate but equal" sixth military branch, the United States Space Force, on 20 December 2019.

EPOCH I

Period of Discovery, 1907–1941

Between 1914 and 1918, airplanes were made of canvas, steel tubes, thin aluminum plating, wood, and wire. Compared to aircraft that would be rapidly developed in the years following the Great War, they were flimsy and unreliable; yet it is important to remember that they were the most technologically advanced pursuit and bombardment aircraft of their day and represented an enormous leap forward from the aircraft of 1914. They were true technological wonders. The combat aircraft flown by the first generation of American airmen were not "American-made," but were built by the French and the British. Initially, these included the Nieuport variants: the 11 (nicknamed the *Bébé*) and the 17. There was also the Sopwith Camel of British design, which lives on in the modern era as the plane in which the Peanuts character Snoopy—as his alter ego the World War I flying ace—takes to the sky against his nemesis, the Red Baron. Five American squadrons also piloted them in the skies over Europe. One of the aviation companies that built them was the Société Pour L'Aviation et ses Dérivés, which American fliers accordingly called "SPADs." The SPAD S.XIII became the most recognized aircraft flown by Americans during the Great War. To this day, the 94th Fighter Squadron, Eddie Rickenbacker's famous "hat-in-the-ring gang," continues to be known as the SPADs.

The engines of the Nieuports and SPADs leaked oil that flew back in the pilots' faces. They had few safety measures; one was a simple lap belt to restrain the pilot. There was no parachute, as that innovation had yet to come into wide acceptance. In some instances, the wings snapped off or collapsed during a dive. Being made of wood and canvas, and filled with gas, the aircraft often caught fire. A pilot's only solution was to slip the aircraft sideways in the hopes of fanning the flames away from him . . . or jump to his death.

Offensively, each aircraft carried forward-firing machine guns that successfully shot through the propellers thanks to a synchronizing gear, itself a step forward from earlier models where pilots either had to mount the machine guns above the propellers, or fly an aircraft large enough for two occupants: the pilot in front and machine-gunner behind. The plane's top speed was only 135 miles per hour, but it could climb to 20,000 feet into the sky. There were no oxygen systems to help the pilots breathe at higher altitudes—at least most pilots preferred not to use the ones available—and no fire-retardant flight suits. These amenities would only become standard later. To protect against the colder temperatures at higher altitudes, a pilots' uniforms typically consisted of "weatherproof coats, goggles, gauntlets, leather boots." Leather was the preferred material.[1]

Many of these planes were garishly painted; the world still recognizes the famous bloodred Fokker Dr.1, flown by Manfred von Richthofen, the Red Baron, again helped along by Snoopy, thanks to Charles Schulz. Most American squadrons took a more subtle route and simply featured a unit designation on the side, which was sometimes followed by a number: there was the hat-in-the-ring of the 94th Aero Squadron, the skeleton and scythe of the 13th Aero Squadron, or the attacking eagle of the 27th Aero Squadron. The numbers allowed other pilots to know at a glance who was flying which particular aircraft. There existed no way to communicate between aircraft other than hand signals or a rock of the wings.

These were the aircraft American observation and pursuit pilots flew from 1915 to 1918. Once the United States joined the conflict in an official capacity, American aviators flew bombardment missions as well. Together, they helped create the aerial way of war of the first epoch by defining the roles and missions an "air force" was capable of doing. They also created, for all future generations of American airmen, an independent and unique culture of "air-minded" individuals. After the First World War, they argued vehemently for their roles and missions, but also for the very control of American airpower. If the United States air arm, such as it was, had a way of war before 1941, it was that airplanes contributed to the war effort, though no one was quite sure exactly how.

CHAPTER 1

World War I and the Birth of American Airpower

> Without indulging in prophecy, nothing is more certain than that the era of war in the air is upon us.
>
> —*Maj. Gen. George O. Squier*

When the United States entered the First World War in 1917, the nation's air force, the United States Army Air Service, was woefully unprepared for combat. Future airmen shared this sentiment. The air arm of the American Expeditionary Force (AEF) was green and ill-equipped to face an enemy that had been engaged in combat for the better part of three years. On the bright side, the new American air arm had a lot to rely on. The Allied Powers, involved in the war much longer, provided excellent leaders and training programs. Moreover, veteran American pursuit pilots from the Lafayette Escadrille and the Lafayette Flying Corps, along with other American volunteer units, flowed into the new squadrons to provide leadership and expertise on how to survive in the sky. The major question to be asked is, What did the United States Army Air Service hope to accomplish during the war? To answer that question, one must first look at the development of the American air arm.

The very idea of aerial combat, as it emerged during the First World War, was suited to uniquely American attributes, or at least attributes that some considered uniquely "American." An early history of the war stated that American fliers found themselves drawn to flying for "the novelty, the picturesqueness, or the fascination of the air-ship service." This sentiment should be juxtaposed against the hard truth of air combat: a short life expectancy and a likely death alone above the clouds—a life that was far from novel and picturesque.[1]

The genesis of the American air arm occurred only a decade before some Americans found themselves engaged in aerial combat for the first time. America's interest in aeronautics prior to the First World War was episodic at best, and this reflected the attitudes of US government and Army leaders as well. Samuel Pierpont Langley acquired a grant from the military for a flying machine, but the military's interest waned after Langley's machine plummeted into the Potomac River in 1903. Later that year, the Wright brothers flew at Kitty Hawk, but the military showed no appreciable interest in the achievement or in purchasing an aircraft for official use, until, that is, the Wrights began selling their designs in Europe.[2]

Prewar Air Arm: Airships and Airplanes and Signal Corps No. 1, 1907–1911

America's military air arm officially began when it was activated on 1 August 1907 as the Aeronautical Division, United States Army Signal Corps, United States Army. Ironically, the Aeronautical Division did not actually have any aircraft in its inventory at the time. The US Army Signal Corps held a competition looking for individuals or a company that could build an airplane with a range of 125 miles, a minimum speed of 40 miles per hour, and the ability to remain aloft for one hour while carrying two people. It took another year for the Signal Corps to decide that the obvious pair to build the first American military aircraft were the Wright brothers. The gentlemen from Dayton, Ohio, provided the first American military airplane, the "Signal Corps No. 1," at a cost of $30,000.

The Wright brothers delivered the first airplane to the Army Signal Corps at Fort Myer, Virginia, in August 1908. A month later, on 17 September, Lt. Thomas E. Selfridge, flying in the aircraft with Orville Wright, became the first aviation casualty of the US military when the propeller broke, severing a guy wire, which resulted in a crash. Selfridge became the first of the early aviation pioneers whose names now adorn airfields and airports across the United States.[3]

The diary of V. L. Burge, a private in the Signal Corps, and an article he wrote titled "Early History of Army Aviation" are instructive of what life was like at the first "air force base," such as it was. Burge was on hand when the Wrights arrived with their airplane. He also worked directly with Lts. Frank Lahm and Benjamin Foulois, two of the Army's earliest pilots and both US aviation pioneers. Burge witnessed the visits to the airfield of President Howard Taft and Speaker of the House Joe Cannon. Burge was within earshot when

Cannon remarked, "No-one can convince me such a thing will fly," and, after the Wright Flyer did leave the ground, remarked again, "Well, it's in the air, but you can't make me believe it will stay there."[4]

According to Burge, it was Foulois who decided to install wheels on the Army's Wright Flyer, which speaks to the primitive nature of aircraft at the time and the necessity of innovation as early fliers improved their aircraft through trial and error. Where Burge's story ends, Foulois's begins. Foulois learned to fly with the Wright brothers and became one of the military's first pilots. (Debate continues about who may actually claim the "first" title, the three leading candidates being Foulois, Frank Lahm, and Frederick E. Humphreys.) Many decades later, Carl Spaatz sat down with both Foulois and Lahm in the "interrogation room" at Maxwell Air Force Base where they were sitting together with two bottles of whiskey. "Tooey," as Spaatz was known, poured drinks for both Lahm and Foulois and took a big one for himself saying, "Who really was the first military aviator?" The answer, according to Foulois's autobiography, *From the Wright Brothers to the Astronauts*, is Frederick Humphreys, followed by Lahm and then himself—though Foulois might have been the first to receive the *title* of military pilot, thus allowing him to share first place.[5]

The Punitive Expedition

In March 1916, while the war in Europe raged on, having already claimed millions of lives, the United States air arm joined Brig. Gen. John J. Pershing's Punitive Expedition into Mexico against Francisco "Pancho" Villa. An official Air Force history, written years later, stated that this represented the first time an air unit was used as a tactical asset in the field, but added, somewhat sardonically, "It was in for a rough time."[6]

The US Army expedition sent to find and capture Villa came about after his insurgent forces attacked and burned the border city of Columbus, New Mexico, a raid that left ten civilians and eight soldiers dead and outraged Americans across the country. Chief of Staff of the Army Maj. Gen. Hugh Scott ordered Pershing to track Villa down.[7]

The US Army's 1st Aero Squadron, commanded by Captain Foulois and composed of eight JN-3 "Jennys," traveled south by rail to provide aerial observation and reconnaissance. With the 1st Aero Squadron was twenty-three-year-old Lt. Carl Andrew Spaatz. Spaatz's biographer, David R. Mets, noted in *Master of Airpower* that the squadron spent "a fruitless month" in Mexico before being ordered back to the United States. Spaatz, however, looked back on the experience fondly, remembering in 1968 that the crews were young, inexperienced, and "somewhat irresponsible," but they enjoyed using airplanes to contribute to the expedition.[8]

Foulois's remembrances of the expedition were not so halcyon, and a look at the data bears him out. It was an inauspicious start for America's air arm. Of the eight planes, two were lost early on—one of these, after the pilot ditched in the desert and escaped into the mountains, was destroyed by Villa's men, making it the first American aircraft destroyed by enemy forces. Foulois himself later crash-landed and was "captured" by the population of a local village, making him, in his own words, "the first United Sates aviator ever to become a prisoner of war."[9]

Even when Foulois and his men could get the aircraft aloft and remain in the sky, they were unsuccessful at providing timely and relevant observation or reconnaissance. Of the remaining six aircraft, none were advanced enough to conduct operations in the mountainous region known for its high winds. Foulois was forced to report to Pershing that his detachment was "not capable of meeting present military service conditions."[10]

By the time the squadron shipped back to the United States, only two of the original eight aircraft remained in serviceable condition. The Punitive Expedition proved a less than heroic inaugural mission of America's flying arm. Foulois stated, "The machines were inadequate for the task assigned. Not only were they inadequate, they were downright dangerous to fly because of their age." Still, aircraft manufacturers back in the United States were building new aircraft and adapting the red-designated JN-4s to other purposes and roles. The Jenny would remain in service throughout the 1920s, becoming the preeminent aircraft in the years following the First World War both for training American pilots and in the burgeoning civilian aviation industry. However, across the ocean on the battlefields of Europe, aircraft were being developed rapidly in the form of fighters and bombers. The American aircraft industry entered production too late to provide adequate numbers of aircraft for American units.[11]

What did this all mean for America's air service? The simple answer is not very much. It was difficult enough to merely get the aircraft flying, much less to conduct the observation and reconnaissance missions assigned. Still, every journey begins with a first step, no matter how inauspicious, and in short order the United States Army Air Service got its chance to prove itself in combat, just not in American-made aircraft.

The Role of the Airplane in World War I

It would be years before Americans—outside of some units of independent aviators—entered the fray in Europe. By that time, the role of the airplane in combat had developed into defined missions. In the First World War, the airplane emerged first as a means of reconnaissance and observation and only later as an actual instrument of combat. At the outbreak of hostilities, the role of aircraft focused entirely on artillery spotting, observation, and reconnaissance; only as belligerents sought to stop the other side from conducting reconnaissance did air-to-air combat emerge, followed shortly by aerial bombardment. The Zeppelin raids played a part as well. As German forces moved rapidly through Belgium and France, they became separated from each other, the deeper into the French countryside they moved. Airpower played a defining role here, aiding the French in their race to the Marne River to exploit a gap in the German lines. After an observation aircraft reported the gap, French forces were able to rush from Paris to smash into the Germans, leading to the Battle of the Marne, which halted Germany's push into France.

More broadly, each of the combatants across Europe continued to use aircraft as a reliable source of reconnaissance. Obviously, each side also wanted to prevent their opponents from flying over their own lines, and both sides attempted to interdict the other, first with pistols and shotguns and later with mounted machine guns. It was the birth of air-to-air combat. Roles and missions developed rapidly for the burgeoning air service of each nation, as did the understanding of the capacities and uses of the airplane in combat. Two roles merit discussion here as they pertain to the development of airpower in the First World War: pursuit and bombardment.

Pursuit

It did not take long after the Battle of the Marne for combatants on both sides to see the utility of airplanes beyond a reconnaissance role. As a result, a new scourge arrived over the skies of France: the pursuit pilot—what today is known as the fighter pilot. The purpose of the pursuit pilot was to engage in aerial combat and destroy the enemy's aircraft or balloons. Such pursuit could be offensive— flying over enemy lines—or defensive—flying patrols over your own lines. A pursuit unit might play a protective role for reconnaissance aircraft or gain air superiority over a ground offensive. Pursuit pilots rapidly gained the attention of the media and then the public. One such aviator, in particular, Manfred von Richthofen, was soon making a name for himself. Although Richthofen certainly became the most famous of the World War I flying aces, other names, in other countries, also became ensconced in history: René Fonck of France, William "Billy" Bishop of Canada, Edward Mannock of the United Kingdom, and Eddie Rickenbacker of the United States.

With the centenary of World War I having ended in November 2018, these names slip further into history and further from recognition. With eighty kills, Richthofen remains unarguably the most effective of the World War I aces (an ace being an individual with more than five aerial kills). The term "ace" as applied loosely by all combatants was created for propaganda uses and morale-building purposes. Of course, Richthofen's fame is due not only to his status as the "ace of aces" in the war, but to a prolific posthumous career "fighting" Snoopy, marketing Red Baron pizzas, and being the subject of a 1960s song by the Royal Guardsmen. Richthofen credited his success to the German ace Oswald Boelcke, whose famous "Dicta," or rules for fighting in the air, became a mainstay of pilot education in many countries after the war. Boelcke's Dicta stated the following:

> Always try to secure advantages before attacking. Climb before and during the approach in order to surprise the enemy from above, and dive on him swiftly from the rear when the moment to attack is at hand.
> Try to place yourself between the sun and the enemy. This puts the glare of the sun in the enemy's eyes and makes it difficult to see you and impossible for him to shoot with any accuracy.
> Do not fire the machine guns until the enemy is within range and you have him squarely within your sights.
> Attack when the enemy least expects it or when he is pre-occupied with other duties such as observation, photography, or bombing.
> Never turn your back and try to run away from an enemy fighter. If you are surprised by an attack on your tail, turn and face the enemy with your guns.
> Keep your eye on the enemy and do not let him deceive you with tricks. If your opponent appears damaged follow him down until he crashes to be sure he is not faking.
> Foolish acts of bravery only bring death.
> The Jasta [squadron] must fight as a unit with close teamwork between all pilots. The signals of its leaders must be obeyed.
> For the Staffel [also squadron]: Attack on principle in groups of four or six. When the fight breaks up into a series of single combats, take care that several do not go for one opponent.[12]

One of these rules, "Try to place yourself between the sun and the enemy," created an Allied response: "Beware the Hun in the sun."[13]

These sets of rules were distributed to younger pilots and new recruits joining pursuit units, and they provided the initial lessons from a man they all looked up to. But Boelcke was not the only one creating and disseminating aerial doctrine. In the Royal Flying Corps, James McCudden, a British ace and an engineer, helped codify how to be a fighter pilot in the pamphlets *Fighting in the Air* and *Notes on Aeroplane Fighting in Single-Seater Scouts*. In Germany, the United Kingdom, and France, pilots collected, discussed, and distributed rules for combat in the air. Later called "tactics, techniques, and procedures," these rules helped new pilots learn the way of war of their organization. The same would be true as American pilots joined in the fray.[14]

Bombardment

Rarely in the history of warfare or technological development has one instrument accomplished so little and had so much of an impact. The concept was simple enough: aircraft overflying the front lines could strike at the enemy's rear by moving beyond the forward edge of the battle area directly to staging areas, jumping-off points, or even further back to the actual manufacturing centers of the enemy. Rapidly, proponents realized these attacks could take two forms: bombing to destroy industrial targets and bombing to destroy the morale of the citizens and the soldiers. The First World War engendered what historian Craig Morris called a "theoretical awakening," with regard to strategic bombing.[15]

Prior to the war, the concept of any type of airship—lighter- or heavier-than-air—overflying a city and unleashing weapons on an unsuspecting populace was found more in the works of H. G. Wells and Jules Verne than in accepted military theory. *The War in the Air* (1908) by Wells depicted Imperial Germany launching air raids against the United States. Ironically, the book states that it was America that posed the greatest threat to Germany's aerial ambitions as the only nation that could become a rival in the air. The actual events of World War I did not bear this out.[16]

Airpower theorists were just that: those inventing theories of what the airplane or airship might be able to accomplish. The Italian theorist Giulio Douhet began writing about strategic bombardment around 1908. In January 1915, the Germans were the first to use large airships for bombardment and to achieve psychological purposes with the Zeppelin raids against London. These were followed later in the war by the Gotha heavy bombers. The impact of these raids endured for decades and led directly to a British response, headed by Lord Tiverton, 2nd Earl of Halsbury. Tiverton's theory became, if not synonymous with, then certainly emblematic of early American bombardment theory: it defined target sets and created accompanying lists to be attacked as part of an "industrial web" process. Lord Tiverton, though forgotten to all but the most ardent of airpower scholars, holds an important place in the development of American strategic bombardment. Air Corps students, graduates, and instructors at the Air Corps Tactical School later used Tiverton's ideas and processes to develop the American air plan for the war in Europe prior to the Second World War.[17]

Douhet and Tiverton's ideas, the literature of Wells and Verne, and the Zeppelin and Gotha raids all played into American conceptions of strategic bombing and produced a lasting influence on American doctrine after the war. Billy Mitchell was the foremost proponent of strategic bombing—and the most vocal—but Edgar Gorrell and Benny Foulois also contributed greatly to the direction American air war theory took after the war. As will be seen, two forms of aerial combat—pursuit and bombardment—laid the foundations of American interpretations of airpower.

The Lafayette Escadrille (Flying Corps)

Although it would be nearly three years after the start of the First World War before an American squadron from the AEF made its presence felt, a contingent of American pilots sought to get into the war in the air earlier, even if that meant fighting in the British or French air forces. This group of men learned the art of aerial warfare and later would provide their experiences and tutelage to America's airmen in the AEF. Forever associated with the name Lafayette, they created the first concrete version of American airpower. However, it should be noted that while the French Layette Escadrille was certainly the best-known unit composed of American fliers, it would be disingenuous not to mention those American airmen who flew in British units prior to America's entry into the war. Despite this, it was the survivors of the Escadrille who went on to form the leadership of later all-American units, and thus secure their place in history.

By the time the United States declared war on Germany in April 1917, the level of matériel and monetary investment the United States had made in the development of some type of "air force" made it only the fourteenth-largest investor in the world in terms of air force expenditures, behind not only the European powers, but also Japan, China, Greece, and Brazil. To say America was behind would be a gross understatement. As Juliette Hennessy's *United States Army Air Arm, April 1861–April 1917*, an official Air Force history, phrased it, "At least one thing is certain: At the outbreak of war little or nothing was on hand, either of planes, fields, instructors, curricula, or—most important of all—experience that would indicate what was needed."[18]

For this reason, American pilots flew British and French aircraft into the air during the war. However, due in no small part to the earlier work of the Lafayette Escadrille, American units had a hardened cadre of officers and maintainers, or mechanics, who infused their squadrons with continuity, combat experience, and an institutional know-how that they passed on to the new American fliers. This experience and expertise manifested itself in how to *kill* the enemy and at the same time not *be killed* by the enemy. As Boelcke and Richthofen instructed the Germans, the survivors of the Lafayette Escadrille did the same for the new American pilots. The early pursuit pilots and bomber units that were part of the formal American air arm therefore entered into aerial combat on the wings of experienced combat leaders. The cumulative experience of American fliers in the First World War gave birth to an aerial way of war that is still in evidence in American fighter units in the twenty-first century.

The American experience in the air culminated in the Battle of Saint-Mihiel, under the command of Brig. Gen. William Mitchell. In this the largest aerial battle the world had ever seen, more than 1,400 aircraft of all types took to the skies.[19]

For the American aviators and support personnel flowing into Europe in 1918, the members of the Lafayette Escadrille provided a silver lining to the otherwise untried and untested pilot corps. With their presence, there already existed an all-American unit that had everything the air arm was short of, but above all experience in combat. Since 1916, American fliers had been participating in the war in the air. Some of these men were already aces, and all were combat veterans. The men of the Escadrille and the later Lafayette Flying Corps knew the enemy, and they knew his tactics. In the book *America in the War: The Vanguard of American Volunteers in the Fighting Lines and in Humanitarian Services*, published shortly after the war ended, Edwin Morse stated: "No historian of the future will be able to ignore the important part which that small but heroic band, the Vanguard of American Volunteers, played in the great war to make the world safe for democracy.... A handful of the more daring spirits entered the French flying corps and formed the nucleus of what later was to become the Lafayette Escadrille."[20]

The idea of an American squadron flying under French command was the brainchild of Norman Prince. A 1908 graduate of Harvard and two years later of Harvard Law, Prince conceived of the "Escuadrille Americaine." At the time, there was no "room at the inn" for Americans wishing to join France's aviation branch. The only way for an American to join the war effort against Germany was through the French Foreign Legion, even though some Americans living in France at the time, including Prince, Elliott Cowdin, and William Thaw, already had extensive flying experience. Thaw, a member of the class of 1915 at Yale, traveled to France in 1914 and joined the French Foreign Legion. In November of that year, Thaw transferred to the French Aviation Service, citing his pilot license, but instead of flying the aircraft, Thaw found himself in the rear as an observer. Still, it was better than the Legion. Thaw's perseverance paid off, as he eventually passed all of his tests to become a pilot and was eventually sent out to join a French escadrille.[21]

During the Christmas season 1915, Prince, Cowdin, and Thaw returned to the United States, and later they eventually returned to France together. On the return trip to Paris, Prince broached the subject of an all-American flying squadron. Cowdin and Thaw agreed. The three men then enlisted the help of Georges Thenault, a French officer and pilot, to navigate the myriad French government offices from which they needed to gain approval. Thenault later recalled that Prince was rabid in his desire for an American flying squadron. According the Thenault, Prince "declined to recognize difficulties. I had to calm him down or he'd have sent an ultimatum then and there to all the French authorities." Edwin Morse connected the Americans' desire to fly directly back to the perceived American values of the age. Flying in France offered "free play in a limitless medium to individual initiative, judgment and skill. This was a form of warfare which harmonized perfectly with American traditions and with American temperament." However, this glorifying statement obfuscates the realities of war in the air.[22]

The all-American unit officially came into being on 21 March 1916, with the support of the French Air Department, and entered into combat less than two months later. In April, orders were sent to any American fliers in French air units to join Escadrille N 124—the numeric designator that came to be known in French as the "Escadrille La Fayette"—at Luxeuil-les-Bains (air base). The initial cadre of American fliers included Prince, Cowdin, and Thaw, along with Kiffin Rockwell, Victor Chapman, James McConnell, Bert Hall, and Raoul Lufbery. The ranks of the original members eventually swelled to thirty-eight and included five French officers. On 20 April 1916, these original members of the Escadrille arrived at Luxeuil. The aforementioned French pilot Capt. Georges Thenault commanded the new escadrille. Unlike so many of his subordinates, Thenault survived the war, and in 1921 he published one of the great first-person accounts of the war: *The Story of the Lafayette Escadrille*.[23]

Of this small core of American fliers, perhaps none displayed the "Americanness" the French perceived more than Bert Hall. Claiming to be a pilot, Hall was given the chance to prove his skills. In reality, it was his first time up close to an airplane. Hall clambered into the cockpit, as Thenault remembered the occasion, then "off he went zig-zagging like a drunken duck, actually left the ground, but crashed headlong into the wall of a hangar." He destroyed the aircraft, but the French decided there was something to his willingness to get himself killed simply to prove himself and Hall continued on to pilot training.[24]

Hall was the embodiment of this early group of Americans: "No matter what happened, no matter where it might be, they wanted to fight"—perhaps too much so. Two other Americans in France, displaying the brashness the French found so annoying in Americans, also claimed they were pilots only to climb into aircraft to prove otherwise. One crashed without taking off, while the other was able to get his aircraft to a height of over 1,000 feet before crashing headlong back to Earth, but he somehow survived. Both men were allowed to continue with flight training.[25]

At Luxeuil, the Americans began their training on the Nieuport 11, known as the *Bébé*, or "baby Nieuport." The Bébé carried one Lewis machine gun mounted below the upper wing to allow it to fire over the propeller (the synchronizer that allowed for a machine gun to fire through the propeller was not in use in this early model). It had a maximum speed in the range of 95 mph. From the Bébé they moved on to the more modern pursuit Nieuports, the *avion de chasse*.

The Escadrille's first victory came on May 18 when Kiffin Rockwell located a German Luft-Verkehrs-Gesellschaft (LVG) reconnaissance plane and brought it down by diving on it and firing a quick burst. One Air Force history described Rockwell's method of attack and victory this way: "A veteran hunter or more cautious pilot might have seized the opportunity to surprise the LVG and launch an attack out of the sun or from behind a cloud, but this one approached directly, without guile.... The chasse pilot (Rockwell) closed to point blank range and, just as a collision appeared imminent, fired a quick burst, then swerved away. The encounter was over that quickly. Both the observer and pilot collapsed; the LVG rolled and plunged to earth; the Nieuport banked away leaving a plume of smoke to mark the scene of combat."[26] Rockwell's brother mailed him a bottle of eighty-year-old whiskey, which the group decided would be reserved for those "pilots who brought down Boches [derogatory slang for Germans]." Shortly after the first victory, the group received orders to travel to Verdun.[27]

At Verdun, Thaw made his first kill. Back on the ground, he told his fellow pilots, "No credit to me. I just murdered him. He never saw me." Thaw's comments were emblematic of aerial warfare over France. While films later displayed swirling aerial dogfights—and this absolutely did occur—most air-to-air kills came when one combatant was able to approach at close range behind an unsuspecting opponent and kill him from behind. A few days later, Thaw himself received a bullet to his arm that forced him to abandon a dogfight and land.[28]

On 24 May, new pilots began arriving to bolster the Lafayette Escadrille's numbers. This new group included Raoul Lufbery. Thenault described Lufbery as "simple, modest, silent, and hardworking." Born in France to a French mother and an American father, Lufbery, by the time he joined the Escadrille, had already spent time in the American Army in the Philippines and the French Foreign Legion. An experienced aircraft mechanic, he differed greatly from the wealthier Ivy Leaguers, but he quickly earned their respect. Despite being known as quiet, Lufbery also had a temper. On one occasion, Thenault received a telegram from Lufbery when the latter was on leave. It read: "Am held in prison at Chartres." Lufbery had gotten into an argument with a railroad employee, and when the worker had grabbed Lufbery's uniform, Raoul knocked him out with a single blow.[29]

There is no doubt the American contingent made their mark in the arena of air-to-air combat. Thaw and Lufbery both became aces. Hall, Prince, and Rockwell had each shot down enemy airplanes. Their leader, Georges Thenault, later recalled of the early days of the organization: "Those were the heroic days of the Escadrille, its glorious prime. Prince, Lufbery, Rockwell, and Chapman, were you not worthy rivals of the greatest Heroes of any age or country?" American historian Bert Frandsen noted that members of the Lafayette Escadrille, due to their daring and courage, and to their exploits being presented in the newspapers, became heroes on both sides of the Atlantic. They received letters of appreciation and children were named in their honor.[30]

Despite the accomplishments of these airmen in the aerial arena, or perhaps because of these accomplishments, the Lafayette Escadrille suffered a very high casualty rate among its initial cadre of fliers. The first to fall was Victor Chapman. Ironically, Chapman had already survived a dogfight where he narrowly escaped death when a bullet grazed his head. On the morning of his death, Chapman loaded oranges into his plane intending to take them to a fellow aviator, Clyde Balsley, who had been injured and was recuperating in a nearby hospital. On the way, Chapman saw a group of Fokkers and "couldn't resist the temptation of attacking the foe who had brought Balsley down." It was, in hindsight, a particularly bad decision, as Chapman engaged in one-on-five combat. But Chapman's penchant for such a fight was emblematic of the spirit of the Escadrille and the willingness of the American airmen to protect, or exact revenge for, their comrades. Nevertheless, Victor Chapman was shot down. French writer André Chevrillon stated, "The death of Victor Chapman touches our imagination with fire." The French ambassador to the United States, Jean Jules Jusserand, stated, "Never in my country will the American volunteers of the Great War be forgotten."[31]

Kiffin Rockwell died on 23 September 1916 at the age of twenty-four. He dove against a two-man German Albatross, but the German gunner scored a fatal shot to Rockwell's head. Thenault lamented later that Rockwell did not attempt to turn in behind the German aircraft or approach from a more advantageous position and instead simply dove directly at the enemy in a perfect field of fire for the German gunner. Lufbery, forced to land earlier, witnessed his friend's death from the ground, helpless to do anything but watch the drama unfold. Rockwell's own words provide a fitting epitaph: "I pay my part for Lafayette and Rochambeau."[32]

Norman Prince, returning from a combat mission as night fell, struck power lines and capsized his aircraft into a field. Initially it was believed he had only broken his legs, but his condition grew steadily worse from a blood clot in his brain. He died on 15 October 1916 at the age of twenty-nine.[33]

One final member of the original Lafayette Escadrille is worth mentioning here: James Rogers McConnell. McConnell suffered a serious back injury in the late summer of 1916, enough to prevent him from flying and land him in a hospital ward. McConnell took time during this convalescence leave to write about his experiences. McConnell's *Flying for France* remains one of the great aviator memoirs ever published. While it is often said that a pilot's odds of being killed in combat can be measured in days or even hours, the chances of being wounded in the air and surviving are often overlooked. In one passage McConnell recounts, "Thaw was wounded in the arm, and an explosive bullet detonating on Rockwell's wind-shield tore several gashes in his face. . . . A week or so later Chapman was wounded." Chapman and Rockwell both died later in combat. Another flier, Clyde Balsley, was hit by an exploding bullet in the thigh. Death lurked on every mission. Experiences like these only increased the fatalism the American pilots lived with. Like a German bullet, it often came close and missed, but death was more than an eventuality.[34]

McConnell's memoir also demonstrates that American pilots considered the humanity of their enemy, viewing them not just as targets to be brought down: "It occurred to me that he might have been making his first flight over the lines, doubtless full of enthusiasm about his career. Perhaps dreaming of an Iron Cross and his Gretchen, he took a chance—and then swift death and a grave in the shell strewn soil of Douaumont." McConnell also notes that he felt more pride in being a sergeant who flew than in being a lieutenant on the ground. *Flying for France* was published posthumously, as McConnell was killed on 19 March 1917. In a letter dated two days before his death, he said, "The season for flying has now opened." It proved to be the last season of his life.[35]

A report put together after the war places the number of members of the Lafayette Escadrille at 43 (the 38 Americans and 5 French officers); 170 Americans served later in the expanded Lafayette Flying Corps. Of these, 11 from the Escadrille were killed and 54 from the Flying Corps. The Escadrille earned a total of 72 aerial victories and the Flying Corps 127. More than a tale of heroism (although none of the members of the Lafayette Escadrille would have called themselves heroic), the major success and importance of the organization would only become evident later. It was not just that Americans proved willing to fight on behalf of France—although that cannot be diminished—but rather that the survivors would go on to play a significant role in educating the airmen of the AEF, as they arrived in theater when America entered the war.

While giving the new Americans someone to look up to and something to strive for, what members of the Lafayette Escadrille accomplished in the long term was making the transition from fighters to teachers and trainers. The surviving members of the Lafayette Escadrille and Lafayette Flying Corps formed a nucleus of combat-experienced Americans who could provide much needed leadership to the pilots arriving in France in the spring of 1918.[36]

Thenault recalled that, thanks to the Lafayette Escadrille, American units received the "benefit of their [combat] experience." Moreover, many of the surviving members became commanders in the new American units. These included James Norman Hall, William Thaw, and Raoul Lufbery. Perhaps the most important of these fliers was Lufbery, whom Rickenbacker later called the "most revered American aviator in France." Rickenbacker biographer W. David Lewis named Lufbery the "legendary dean [of the] Lafayette flyers."[37]

Lufbery fell in combat on 19 May 1918. After taking off from his aerodrome in hot pursuit of German fighters, his aircraft was hit by German fire and burst immediately into flames. Only a few days earlier, someone had asked Lufbery what he would do if his plane caught fire: stay with it or jump? Lufbery said, "I should always stay with the machine. . . . If you jump you certainly haven't got a chance. On the other hand there is always a good chance of side-slipping your aeroplane down in such a way that you fan the flames away from yourself and the wings. Perhaps you can even put the fire out before you reach the ground. It has been done. Me for staying with the old 'bus, every time!" In this instance, witnesses saw Lufbery as he sideslipped the aircraft toward Earth before it became too unbearable and he jumped from the aircraft. As the fight took place almost directly above the airfield, the pilots of the 94th Aero Squadron went to collect Lufbery's body. They found him laid in a local town hall "entirely covered with flowers from the near-by gardens." Lufbery represented the very best the American Air Service had to offer during the First World War. Eddie Rickenbacker, a veteran of the Lafayette Escadrille and a great pilot and ace in his own right, credited Lufbery with teaching him everything he needed to know about flying pursuit. Rickenbacker also called him "the greatest pilot of them all." Lufbery also proved to be a very capable trainer. In his wake and thanks to his careful training of subordinates, he helped create a new predatory animal: the American fighter pilot.[38]

In total, the United States produced over 150 air aces throughout the entirety of the First World War—pilots who scored five or more aerial kills on enemy airplanes or balloons. Of these, 17 served in one of the French escadrilles, before the arrival of the American Air Service in 1918; another 40 served in British units. Perhaps most important, of those airmen who flew with British and French units, fourteen escadrille pilots and twelve from the Royal Flying

Corps would later serve in an American aero squadron and represented an authoritative source of information and training aid for the new American airmen arriving in France.[39]

Arrival of the Americans

Americans had fought in French and British units since 1916, but the official arrival of Americans in Europe did not occur until 1917. And the bulk of American combat forces did not begin arriving on the continent under the banner of the American Expeditionary Force until the following year. It was around this time that a particular American air officer began making his presence felt. The son of a former US senator from Wisconsin, Col. William "Billy" Mitchell was just launching his lifelong quest to prove the efficacy of airpower. His career would eventually end in a court-martial for insubordination, but for the present, he was just one of many army colonels arriving in France with the rest of the American forces.

Mitchell had been sent to the Aviation Section of the Signal Corps in 1916, according to Foulois, to "instill old-fashioned military discipline among the so-called prima-donna pilots then on active duty." Foulois also mused, "This was to me an extraordinary and amusing reason, and, as it turned out, a supreme irony which almost wrecked military aviation in this country." It has been well documented that a great antipathy existed between Mitchell and Foulois, but in light of Mitchell's future actions, Foulois was correct in his assessment. It was indeed ironic that Mitchell's superiors expected him to instill military discipline among fliers, as Mitchell eventually became the ultimate maverick and prima donna to those outside the Air Service. Upon arriving in France in April 1917, Mitchell temporarily replaced Col. Samuel Reber as chief of the Aviation Section (the overall leader of American airmen in the Army), but this was a short-lived assignment as Maj. Gen. George O. Squier was assigned the permanent post. However, Squier wanted someone in France to observe aviation matters, and he chose to send Mitchell to the front.[40]

Mitchell arrived in Paris on 10 April 1917, and spent the next two months observing ongoing aerial operations until Pershing arrived in June and ended his academic studies. It was time to put the AEF to work. Mitchell's strengths lay in his organizational ability and in the loyalty he garnered from his subordinates. Mitchell's leadership qualities when in a position of authority were never in doubt. He was indeed a leader of men able to get his subordinates to achieve the tasks laid before them. It was when Mitchell was placed under someone else's command that he chafed. In the hierarchical organization of the military, where there is always a superior leader, Mitchell could never bring himself to mold or modify his beliefs to those of his commanding officers.[41]

Throughout his time in Europe, Mitchell sparred and argued with nearly every one of his superiors: 1st Army commander Lt. Gen. Hunter Liggett; the chief of the Air Service, Brig. Gen. William Kenly; Kenly's replacement, Benjamin Foulois; Hugh Drum; and Mason Patrick. Once Foulois, an officer and pilot significantly senior to Mitchell, was named chief of the Air Service, he became a popular target for Mitchell. Throughout 1918, as Mitchell biographer Alfred Hurley noted in his book *Billy Mitchell: Crusader for Air Power*, "he became less and less of a team player and took every opportunity to snipe at Foulois." Another Mitchell biographer, James Cooke, noted, "It never really occurred to Colonel William Mitchell to try and accommodate his colleagues or to admit that men like Foulois or Patrick might indeed be good officers with something to contribute to the growth of the AEF's Air Service."

With Foulois, Patrick, and Mitchell running the American air arm, it was time to see what exactly American airmen could contribute to the war effort. The supreme question that needed answering was, What could American airpower (such as it was) do in the First World War?[42]

There is also little doubt that American pursuit pilots became legendary folk heroes to the American public. The name Rickenbacker became synonymous with daring and courageousness. That being said, World War I saw the beginnings of strategic bombardment, which found fertile soil in the minds of American planners in the wake of the war. In short order, and increasingly over the decades between World War I and World War II, strategic bombardment would subsume pursuit aviation as the reason for the Air Service's existence.

If it seems this book has, thus far, focused more on pursuit than bombardment, it is because that is true. However, even the official history *The U.S. Air Service in World War I* skews in favor of pursuit aviation, allotting seventy pages to the "Tactical History of Pursuit Aviation" and only fourteen to "Tactical History of Day Bombardment." The actual results of American bombardment, or any air arm's bombardment, for that matter, during the First World War were limited in nature. Localized raids in the rear of enemy lines against marshaling facilities or rail yards proved that going over an enemy's front lines to attack rear areas was, in fact, possible. It was this reality that grew into more grandiose theories during the interwar period. During the war, bombardment crews suffered the same heavy casualties that pursuit pilots did.[43]

American airpower could later be proud of its accomplishments in the few months it served on the front lines. The American Expeditionary Force and the Army Air Service acquitted themselves well. From Saint-Mihiel through the Meuse-Argonne campaign, American soldiers and airmen made a valuable contribution toward turning the war irrevocably against the Germans and Austro-Hungarians. The simple fact was that the AEF solved a math problem for the British and the French once Russia removed itself from the war. Arriving in numbers as great as 10,000 per month in the spring and summer of 1918, American soldiers filled the gaps for their war-weary belligerent allies. For the air arms of the British and the French, American fliers filled cockpits—primarily in French- and British-built aircraft.

Combat in the First World War for American airmen, as they quickly learned what their British, French, and German counterparts already knew, was nothing short of ghastly. Although they had soft beds to return to at night and an abundance of alcohol to drown their sorrows and chase their demons away, many pilots turned macabre and accepted death as an inevitability, something the Lafayette Escadrille and Flying Corps veterans knew all too well. Of the many first-person accounts written by survivors of the First World War, one stands above all the others: Eddie Rickenbacker's *Fighting the Flying Circus*. Rickenbacker's account demonstrated the horror and difficulties of early aerial combat. During one melee, he recalled, "I saw that a general fight was on between the remaining ten Fokkers and the eight Spads." One of the American pilots crashed headlong into a German aircraft: "It was a horrible yet thrilling sight. The two machines actually telescoped each other, so violent was the impact. Wings went through wings and at first glance both the Fokker and Spad seemed to disintegrate. Fragments filled the air for a moment, then the two broken fuselages, bound together by the terrific collision, fell swiftly down and landed in one heap on the bank of the Meuse!"[44]

The inevitability of death pervaded the consciousness of American pilots; yet their willingness to climb into these machines and fly to the front and engage with the enemy became a hallmark of the American airman in the First World War and a trait that followed in generations of airmen.

Battle of Saint-Mihiel

In the pantheon of American aerial battles, few rank higher in significance and in historical memory than the Battle of Saint-Mihiel. Over one hundred years after the event, it can still be found ensconced in commissioning curriculum and professional military education courses across the US Air Force—and for good reason. The Battle of Saint-Mihiel was to be the greatest concentration of aerial firepower on the western front. Pershing's air commander for the battle, Col. Billy Mitchell, had at his disposal 701 pursuit fighters, 366 observation aircraft, and 414 bombardment aircraft of both the day and night variety. These 1,481 aircraft were concentrated against a narrow salient in the German lines. Mitchell's plan for the battle was simple enough and would set the tone for future air operations: gain air superiority, then strike targets behind the front lines, including supply trains and troops moving forward. Once effective communication could be established between the ground and air arms, pilots could proceed with cover for troops on the ground or close air support, as it is known now. It should be noted that German resistance on the ground was not going to be fierce. The German military had already decided to pull out from the salient, but in terms of planning, organization, preparation, and execution, Saint-Mihiel rightly served as an excellent "map problem" or training exercise for generations to come.[45]

There was no doubt Saint-Mihiel was a victory both militarily and psychologically for the American air and ground units. Pershing personally praised Mitchell's efforts and put him in both for promotion to brigadier general and for the Distinguished Service Cross.

Mason Patrick said of the battle that "despite handicaps of weather and inexperience, the Air Service contributed all in its power to the success of this St. Mihiel operation. . . . The hostile air forces were beaten back whenever they could be attacked, the rear areas were watched, photographed, and bombed. Our airplanes participating in the battle, by material damage and confusion which they caused, helped to increase total prisoners."[46]

Not everyone agreed with Patrick's and Mitchell's assessments. Benjamin Foulois said later, "The air operation at Saint-Mihiel was encouraging but has been exaggerated over the years, mostly by Mitchell himself and writers who have taken his 'history' as gospel." (It is important to note, again, that neither Mitchell nor Foulois particularly cared for one another.) Still others criticized the planning and execution, and one Mitchell biographer said that Mitchell later "grossly exaggerated what had happened at St. Mihiel."[47]

American pilots performed admirably during the battle. Strafing of enemy positions, destruction of enemy observation balloons, bombardment (both day and night), and aerial combat all aided the decisive victory on the ground. Even if an Allied victory was something of a forgone conclusion—the German army was already retreating out of the salient—it was an excellent opening round for the all-American air arm. Retrospectively, Saint-Mihiel has taken on a somewhat aggrandized version vis-à-vis the reality of the battle, but Mitchell's plan, organization of the air arm, and preparations for the battle deserve due credit.

The Meuse-Argonne

If the Germans had already made a predetermined decision to abandon Saint-Mihiel, the same was not true of the Argonne Forest. Beginning on 26 September 1918, it was to be the last major campaign for American forces in the First World War. Having participated in the reduction of Saint-Mihiel and the battles above Château-Thierry, Maj. Gen. Mason Patrick, now head of the Air

Service of the American Expeditionary Force, found the American pilots to be on par with their Allied counterparts. After the war, he pronounced, "Our pursuit, on entering the Argonne-Meuse operation, had reached a stage at which it ranked in efficiency with pursuit aviation of the armies. It now consisted of three groups of highly trained Squadrons with pilots second to none." The same Air Service history also stated, "It was felt at last that the American units then on the front had developed into trained combat organizations."

The first order of business for the American pursuit pilots in the Meuse-Argonne campaign was to gain air superiority over the German fliers above the forest. In this regard, the Meuse-Argonne, like Saint-Mihiel, aligned with later concepts of airpower: no operation on the ground was possible unless the air arm first gained control of the air. In the aerial battle, the Germans attempted to wrest control of the skies from the Americans by sending up more than a dozen aircraft at a time to engage in what one American pilot called a large "rat fight."[48]

The AEF sent three American Pursuit Groups into battle (the 1st, 2nd, and 3rd). The orders were simple enough: "[Penetrate] about 12 kilometers beyond our advancing lines to clean the air of enemy aviation." Among those in the air that day who earned himself an aerial victory was a young major, Carl Spaatz. Two German aircraft nearly downed Spaatz by rolling in on his six position (immediately behind the aircraft's tail), but his eagle-eyed squadron commander Charles Biddle saved Spaatz with a diving attack on the German fighters. Biddle shot down one of the Germans, and the other made a break for more friendly territory. Biddle's efforts were for naught, however. Shortly thereafter, Spaatz ran out of fuel and crashed anyway, though he lived.[49]

Up and down the front line, American bombers and reconnaissance assets took to the sky. American bombers dropped one and a half tons of munitions on a railway and bridge along the Meuse River. However, early on, American reconnaissance flights were hindered not only by the weather but also by the fact that some of the US troops they were supporting advanced directly through the forest, which made it difficult, if not impossible, for the pilots to keep track of the troops on the ground. The difficulty in observing troops on the ground continued throughout the campaign. This has been a perennial problem plaguing pilots attempting to provide support to ground troops: geography matters and is outside the control of those flying overhead.[50]

American bombardment squadrons and groups faced heavy odds as the Germans attempted to interdict them in the air. Of the consequences of bombardment, Chief of the Air Service Patrick noted, "Such is the demoralizing effect of bombing that the enemy in an effort to prevent it will attack with all his available forces and at whatever cost." This would prove to be a prescient statement when one considers the defensive efforts the Germans used to stop Allied bomber formations in the Second World War.[51]

Conclusion

What American airpower did during the First World War is easy enough to quantify. In his "Final Report of Chief of Air Service," Mason Patrick stated that as the war closed, the Air Service contained 45 squadrons and 761 pilots, 481 observers, and the accompanying troops supporting the ground units. The Air Service possessed 740 airplanes. In his report, Patrick already was referring to this organization as "our air forces." American pilots shot down a total of 781 enemy airplanes and 73 balloons (14 of which belonged to the "Balloon Buster" Frank Luke). Americans participated in a total of 150 bombing missions and dropped 275,000 pounds of explosives. Two distinct types of pilots led the air arm: pursuit and bombardment. One or the other would dominate the air arm for the next hundred years. Each developed its own advocates and icons. Each also developed a unique culture with practices and rituals, some of which continue today.[52]

What this all means is harder to say. Historian Lee Kennett stated in his book *The First Air War: 1914–1918* that "while the role of the air weapon in the Great War was a modest one, the role of the Great War in the rise of air power was anything but modest." Aircraft and airpower represented a comparatively new form of warfare. Many nations, particularly the United States, spent the next two decades expounding, and expanding on what airpower could do. What was clear already during the First World War was that specific mission types already existed: pursuit, bombardment, and intelligence gathering through reconnaissance being the most recognizable.[53]

These mission types evolved during what later became known as the interwar period. Into this development stepped the icons of American airpower, primarily men, who helped shape the air arm of the Army and the Air Force for the next fifty years. Very few of them participated in the First World War, Carl Spaatz being a notable exception. However, what they learned from this war shaped the development of the air arm in the intervening decades. These men looked backward to look forward. They extrapolated from the First World War what they wanted to see in any future conflict. As they were doing this, a technological explosion in aircraft design and capabilities occurred.

Many Americans viewed the fliers of the First World War, especially the pursuit pilots, as heroic figures. Rickenbacker and Frank Luke appeared in serials and comics. Rickenbacker, with Clayton Knight, created his own strip titled *Ace Drummond*. Pursuit pilots symbolized a new heroic interpretation of warfare. They were knights of the sky—flying, jousting, and dueling against the enemy. Pursuit and fighter pilots became heroes to their nations, celebrated, read about, and acted out by children. In reality, these pilots lived a life that might best be called short, brutish, and nasty. William Sherman, himself a pilot serving on various staffs during the war, later wrote in his book *Air Warfare*, "Quarter was neither asked nor given.... The customary action of the air fighter seems peculiarly ruthless. It is far from the precepts of chivalry to pursue an obviously crippled and helpless opponent, firing into him steadily, until a fatal crash into the earth or the certain actions of flames in the air, assures his destruction."[54]

Perhaps the pilot had certain accommodations and advantages over the men in the trenches, but any advantages he possessed had to be balanced against a short life expectancy and the likelihood of a death, alone and scared, in the skies. The longer an individual pilot survived in aerial combat, the more comrades he lost—and the more an impending sense of imminent death hung over his life. Those who survived the war, along with their leaders, most notably Mitchell, sought more autonomy in how they ran their service. Mason Patrick and Benjamin Foulois proved to be ideal leaders, but it was Mitchell—antagonist, theorist, propagandist, and soon-to-be subversionist—who surrounded himself with junior officers and veterans of the war that began agitating for independence. Patrick advocated for independence, but without seeking public attention as Mitchell did. Mitchell returned to America with the fervent belief that "the day has passed when armies on the ground or navies on the sea can be the arbiters of a nation's destiny in war. The main power of defense and the power of initiative against an enemy has passed to the air."[55]

Patrick did not gain the notoriety or the disciples that would surround Mitchell everywhere he went. Patrick, no matter his true importance, was but a clergyman to Mitchell's messiah. It was Mitchell who rose to the level of religious martyr to his devout followers. It quickly became apparent to this fraternity of "air men" that "the U.S. Army, the War Department, and Congress never really appreciated or understood air power." No sooner had the First World War ended than the fight for independence of the air arm began in earnest.[56]

CHAPTER 2

The Interwar Period and the Beginning of Independence

> Air power may be defined as the ability to do something in the air.
>
> —William "Billy" Mitchell, Winged Defense

During the First World War, at least in myth, memory, and historical reports, it seemed evident that pursuit was the dominant force in aerial warfare, but in terms of impact on future operations, the emphasis rapidly transitioned to bombardment. If the First World War is misremembered for white scarves in the wind, those who studied it turned to bombardment as the air force most capable of future contributions.

After the First World War, Mitchell noted, "Air Power is applied by means of definite military air units which are organized, armed, equipped and trained for definite and specific purpose. One kind of airplane or air organization can no more perform all the duties required of an air force than can artillery, cavalry, or infantry acting alone form a whole army."[1] Mitchell and other members of the Air Service recognized the need to further develop the doctrines of the war into a functioning air arm capable of operations across a spectrum of mission types, but capable of operating as a cohesive whole.

While this book is ostensibly about combat and a particular way of *war*, it is impossible to link the two world wars without looking at the development of the American air arm in the interwar period. Besides, there was just as much infighting between the Army Air Corps and the other services that perhaps use of the word "war" is warranted. However, as the Air Service was but one component of the US Army, this chapter is primarily concerned with the dissensions and discussions taking place inside the Army. This chapter covers intraservice rivalry between the Air Service and the rest of the Army. It discusses how the Air Service and later Army Air Corps leaders began to think and conduct themselves as separate and different from regular Army officers. It was this streak of "maverickism" that helped establish an identity for the Air Corps separate from that of the regular Army.[2]

While this occurred, of first importance were developments in technology and the pushing of the bounds of the possible allowed the Air Service and Air Corps to accomplish new feats of aerial wonder, including the circumnavigation of the globe and aerial refueling. Second, Billy Mitchell plays an outsize, but admittedly necessary, role in this chapter. While it would be preferable to focus on the inner workings of the actual leaders of the Air Service, in the 1920s it was, and is today, impossible to ignore the role played by Mitchell and his impact on public perception of the Air Service.

Finally, the development of American doctrine at the Air Corps Tactical School demonstrated that while strategic bombardment certainly became the dominant theory, members of the Air Corps did not ignore air superiority (what they called pursuit aviation), close support, or attack aviation. Personalities latched on to these concepts and then utilized them during the Second World War. Their promoters became linked with their preferred method of using military aviation, and their names became synonymous with their preferred approaches: Chennault for pursuit, Kenney for attack, Quesada for close support, and dozens of others for bombardment. This chapter also looks at how technology became intertwined with the Air Force's identity.

The concept of aerial bombardment, while hardly developed entirely out of altruism, was certainly conceived with the intention of saving lives on the ground, most notably the lives of American soldiers. If bombardment aircraft had the ability to destroy factories, railroads, and marshaling yards, then the need for the grinding, attritional, bloodletting type of warfare witnessed between 1914 and 1918 could be rendered irrelevant. Thus, the promoters of aerial bombardment, what came to be called "strategic bombardment," viewed this method of attack as a means to save lives and shorten war. It was this desire to go "over not through" that began shaping American airpower thought and doctrine after the end of the Great War. Aerial bombardment theory was rooted first in the concept of saving lives and shortening war, but it also became wedded almost immediately to the movement for independence of the air arm from the rest of the regular Army. It was a mission *uniquely suited* to American aviators, its advocates argued, and thus was viewed from the perspective that it should be *controlled* by aviators. Airpower, therefore, was seen as inherently apart from the other branches of the army—infantry, artillery, cavalry, armor—and not tied directly to their execution. Airpower's proponents viewed the possible contributions of airpower as separate from combined arms. So while pursuit, bombardment, attack, and close support needed to work together, with the exception of the latter, they did not necessarily need to work with the rest of the US Army.

After the war, the western European nations took a long hard look at what airpower did and what it might accomplish with further advancement and expenditures. Each country came to a different conclusion about the validity of airpower and how the airplane might be placed into existing constructs of warfare. The fundamental question was, What does the airplane do in warfare? Differing answers and opinions emerged. These led to the development of the first air theories and the rise of the first theorists of airpower, such as Giulio Douhet in Italy and William C. Sherman in the United States.

Perhaps the most important work—at least the one with the longest-lasting effect—to come out of the interwar period was Douhet's *Command of the Air*, published in 1921. Douhet wrote much of his book while serving a prison sentence during the war. The first American translation arrived in the United States in the early 1920s and quickly found its way to the Air Corps Tactical School. It was certainly being read by the bombardment instructors, at the least, by the early to mid-thirties, when strategic bombardment theory began to take hold at the school and the young cadre nicknamed the "bomber mafia" was rising to prominence. It is easy to see where those air-minded individuals liked what they found in the pages of *The Command of the Air* and why traditional groups in the Army believed it all to be fantasy. Early in the book, Douhet argued for one thing: the independence of the air arm. As he stated, it was an "illogical concept of utilizing the new aerial weapon as an auxiliary to the army and navy."[3]

Douhet went on to state unequivocally "that to have command of the air is to have victory." Thus, the purpose of airpower was to control the sky, and to have control of the sky meant having control of the war on the ground, which would thus lead to eventual victory. And the only way to achieve all of this was through an independent air force.[4]

Of equal importance to emerging doctrine and concepts in the United States, but significantly lesser known, was Maj. William C. Sherman's *Air Warfare*. Published in 1927, it represented a culminating point in early airpower theory and provided a bridge from the First World War to emerging doctrines that sustained, and became a hallmark of, the interwar period. It also detailed aerial combat tactics. In many ways, *Air Warfare* was much more important than Billy Mitchell's writings, at least insofar as those who employed military aircraft were concerned. Mitchell wrote to the masses; Sherman wrote to the experts.

Sherman served on the initial faculty of the Air Service Tactical School (ASTS, later ACTS). As one of the initial instructors and assistant to the officer in charge, Maj. Thomas DeWitt Milling, Sherman had the responsibility of helping to create the initial course load. Following a short stint on the staff, Sherman was selected to attend the Rules of War commission at The Hague from November 1922 through early 1923. Following his return to the ASTS, he was selected to attend the Command and General Staff College, where he graduated in 1924. He stayed on as a faculty member, and while teaching at CGSC, he coalesced his ideas and experiences into his book *Air Warfare*. A 1928 "Reading Course for Officers" (something of an early recommended reading list for professionals) listed Sherman as required reading. Sherman was also quoted by J. C. Slessor in his book *Air Power and Armies*, the only American officer to be quoted therein, demonstrating that Sherman's work had at least made it across the Atlantic and was being read and discussed in the RAF. With the Air Service Technical School in place and beginning the slow march toward significant contributions to accepted theory and doctrine, the Air Service itself was in a state of transition.[5]

1926 US Army Air Corps Act

For too long the early history of the predecessors of the United States Air Force have been lost to the overarching tale of Billy Mitchell. Unequivocally, Mitchell is important. However, this focus obscures so many of Mitchell's contemporaries. Mitchell's persona dominated contemporary media coverage, and since that time, it has dominated histories as well. Nearly lost to all but the most ardent of airpower scholars are the names Mason Patrick, James Fechet, Benjamin Foulois, and Oscar Westover, each of whom was a chief of the Air Corps—a position never held by Mitchell. It was these men who organized and structured the Air Corps in the interwar years. While it might be true to say that Mitchell's DNA—a streak of maverickism—still lives on in the Air Force, it was the work and the building of the Air Corps by these leaders that truly brought the organization into existence. Foulois, Patrick, and the others wanted a functioning Air Service capable of working with but also independent of the rest of the Army. Mitchell wanted headlines.

Along with the senior air officers, a number of bills, reports, and acts moved aviation—both military and civilian—into the public light in the 1920s. To name and detail each of them is beyond the scope of this history, but together these policies detail that Congress, as well as the Army itself, struggled with just what to call and how to organize the air arm. In 1915, the National Advisory Committee for Aeronautics was established. In 1926, Congress passed both the Civil Aeronautics Act and the Air Commerce Act in response to findings reported by the Morrow Board and the Lampert Committee. This legislation paved the way for the Air Corps Act, which became law on 2 July 1926.[6]

More about organization that a fighting arm, the Air Corps Act dictated that the "the Air Service referred to in that Act and all subsequent Acts of Congress shall be known as the Air Corps." Historians Wesley Frank Craven and James Lea Cate, authors of *The Army Air Forces in World War II*, noted that the Air Corps Act "effected no fundamental innovation. The change in name meant no change in status: the Air Corps was still a combatant branch of the Army with less prestige than the infantry." Although Benny Foulois remembered that the act gave the Air Corps "a tremendous shot in the arm," despite the fact that the "general attitude of the War Department" did not change.[7]

Maj. Gen. Mason Patrick was named the first chief of the Air Corps, a post he retained until his retirement in December 1927. His successor, Maj. Gen. James Fechet, held the position for the next four years. Patrick asked Foulois to be the deputy chief of the organization. Foulois noted that Patrick was the least remembered of the air force pioneers, despite having led both the Air Service and the Air Corps. Foulois lays this neglect at the feet of the "martyr for air power" Billy Mitchell. Mitchell cast a long shadow, enough to overshadow the Army Corps leaders who served as his superiors and ran the corps during the 1920s.[8]

Patrick merits barely a mention in many official Air Force histories including *Makers of the United States Air Force* and *Reflections on Air Force Independence*. Even his own autobiography, *The United States in the Air*, was short on personal details and his interactions with other leaders. Historian Phil Meilinger said, "Although Patrick was a key player at a most important time in American Airpower history, this book sheds little light on anything of importance during that era." That being said, Meilinger goes on to state that Patrick's "low-key style extensive and friendly contacts within the Army hierarchy, and quiet but relentless pressure achieved results." Patrick's biographer, Robert White, noted in *Mason Patrick and the Fight for Air Service Independence*, "Unfortunately, little is remembered of Patrick, who, in retrospect, was responsible for saving a fledging air force from a variety of self-inflicted wounds and many competing and self-serving outside interests." White always cites Mitchell as the principle reason Patrick has been overshadowed. Patrick proved to be the polar opposite of Mitchell, and it was Patrick who helped usher the 1926 Air Corps Act along, thus making him a true pioneer in the independence movement of the United States Air Force rather than a self-aggrandizing publicity hound solely intent on pushing his own personal works forward. White called Patrick a "progressive visionary in his quest to obtain as much autonomy for the service as possible. His was a balanced and successful approach to air power."[9]

Into the late 1920s and early '30s, the General Staff continued to struggle with the balance of power between the regular Army and the Air Corps. After the assistant chief of the Air Corps Brig. Gen. Oscar Westover asked the General Staff for an end strength of over 4,000 aircraft, the senior leaders balked at the idea and interpreted it as a move by the Air Corps to increase its own position without tying it directly to existing war plans. In response, the General Staff, under the leadership of Douglas MacArthur, formed the Drum Board (named for MacArthur's deputy, Maj. Gen. Hugh Drum) in August 1933. The board was made up of five general officers, including Drum and Foulois, and it opposed the recommendations made by Air Corps planners for development and expansion to meet defense needs—at least, defense needs as viewed by the Air Corps.[10]

GHQ Air Force and the Court Martial of Billy Mitchell

A modicum of aerial independence finally went into effect with the creation of General Headquarters Air Force in 1935. This development is one of the least understood aspects in the history of the American Air Force, for it seemed to create two chains of command and a muddled command relationship. There already existed the headquarters of the US Army Air Corps led by the chief of the Air Corps and then there was General Headquarters Air Force. In essence, GHQ Air Force, led by Maj. Gen. Frank M. Andrews (and later Maj. Gen. Delos C. Emmons), commanded the fielded forces of the air arm, what today might be considered the combat air force. The three combat wings of the Army Air Corps reported to Andrews for training and employment. When Maj. Gen. Oscar Westover became chief of the Air Corps, he retained responsibility for staffing and training of Air Corps members, procurement of aircraft, and matters of supply. Looking at it from the perspective of a modern organizational construct, Westover was the "chief of Staff of the Air Force" responsible for organizing, training, and equipping; and Andrews was the "combatant commander" responsible for employment of the force itself.[11]

General Headquarters Air Force came into existence on 31 March 1935. GHQ Air Force was still subordinate to the Air Corps leaders in Washington, despite the fact that it operated as the "owner" of all tactical Air Corps aircraft. If this arrangement seems disjointed to the modern reader, it was. Andrews, as commander of GHQ Air Force, did not see eye to eye with Chief of the Air Corps Westover. Despite any disagreements or enmity the two might have felt, it never rose to the forefront or a public airing of grievances, the way Mitchell's actions did.

Between Mitchell's court-martial and the reorganization of the Air Corps, the milieu of DC politics kept Air Corps leadership busy. That did not mean that those at the tactical level of Air Corps operations spent their time reading the newspapers about what was occurring at the senior echelons, for they were busy making their own headlines. All the while, the Air Service and, after 1926, the US Army Air Corps continued to experiment and push the boundaries of what it was believed airpower might accomplish. This included innovation and experimentation across the service, and it coincided with continued public interest in all things aviation. Historian Joseph Corn called this the "winged gospel" and noted that two generations of Americans "raised aviation to the status of a technological religion. They worshiped the airplane as a mechanical god and expected it to usher in a dazzling future." It was not aerial combat, but it was important work. The airplane also began to move from a novelty to a true instrument of war. While the First World War demonstrated airplanes' potential in the fields of pursuit, bombardment, and reconnaissance, the work done in the interwar years provided a level of maturity to the adolescent Air Service.[12]

In 1924, after 175 days of traveling, members of the Air Service became the first humans to circumnavigate the globe by air. This was no small feat. The Air Service needed not only the men and aircraft to make the attempt, but also the War Department to work with the State Department to gain the diplomatic agreements necessary from dozens of countries to ensure overflight rights and fueling and repair stops. Four aircraft—the *Boston*, *Chicago*, *New Orleans*, and *Seattle*—began their journey on 6 April 1934, and three aircraft landed back in Seattle on 28 September 1924. Their progress was reported publicly to great interest, and the *Air Service Newsletter* kept track of the "World Flyers" throughout their five-month journey. Beyond the circumnavigation itself, they were also the first to fly across the Pacific in the air.[13]

On the first day of 1929, a crew of five aviators including Maj. Carl Spaatz, Capt. Ira Eaker, and Lt. Pete Quesada boarded a Fokker C-2A with a large question mark painted behind the wing on the fuselage and took off into the California sky. They would not touch terra firma for the next 150 hours. It took thirty-seven air-to-air refuelings to make this flight possible. The Air Corps newsletter said the flight demonstrated the Army Air Corps's "penchant for doing things," as opposed, one might assume, to the Army's other branches' disposition for not doing things. Still, doing "things" was a long way from doing something one might construe as being of use to future warfare. What was needed was a location for air-minded individuals to develop their theories and doctrines on warfare apart from the US Army's more traditional, ground-centric schools. Thus was created the Air Corps Service School. Billy Mitchell immediately hijacked the first class for his own purposes just as it was getting off the ground in both a figurative and literal sense.[14]

Each class was to consist of about fifteen members; the first class never finished the program. Mitchell cannibalized it into his 1st Provisional Air Brigade for the purposes of testing airpower against ships in the 1921 tests off the Virginia Capes. The bombing tests that Mitchell put together provide insight about the difference between training and real combat. Newspapers sensationalized Mitchell's tests, claiming that they had "proven" that airpower could sink capital ships. For its part, the United States Navy was incensed. The bombing tests, from the Navy's perspective, proved nothing. The *Ostfriesland*, the USS *Alabama*, and other ships used for the tests were mere hollow shells sitting dead in the water. They were noncrewed sitting ducks. To the Navy, then, dropping a bomb under ideal conditions was not the same as attacking a ship under full steam with a gunnery crew operating antiaircraft guns that would prevent the would-be attacker from even approaching the ship. This is to say nothing of the difficulty in the 1920s of actually finding a ship at sea. Given that Mitchell violated the preconditions the Navy had agreed to for the tests, naval leaders quickly realized that Mitchell was interested less in accurate results than he was in how the media and the American people might view the results. The Navy's reaction and criticism mattered little to Mitchell. For Mitchell, the bombing tests, and more importantly the subsequent press reports, were a resounding success.

The bombs dropped off the Virginia Capes were not the only explosions in Mitchell's life at the time. His personal affairs were derailing his career, and the high society circles that Mitchell insisted on running in knew about it. His carousing, exorbitant spending, rumored affairs, and known drunkenness caused talk. In 1920, after forgetting his wife Caroline's birthday, he had a row with her that led to Mitchell accidentally—probably—shooting her. Husband and wife disagreed as to the course of the events. The bad blood between Billy and Caroline continued for the next two years, ending in a very messy divorce in September 1922, which also affected his professional career. The chaos of Mitchell's personal life forced his boss and director of the Air Service Mason Patrick to get his deputy out of Washington for the good of both the Air Service and Mitchell as well.[15]

Sadly, the divorce and exile did nothing to stem Mitchell's antics. In sending him on an overseas mission, Mason Patrick warned Mitchell that it would "be most unwise to indulge in any discussion or criticism of the Air Service policies of the United States or to make any statement concerning what the future policy will or should be.... Your mission is primarily and exclusively the gathering of information." Mitchell failed to heed Patrick's admonition to be "very discreet."[16]

His trip abroad only reinforced his belief in the superiority of the airplane over the Army's ground units and the US Navy. Mitchell became more outspoken and used his influence and persuasive skills to get his thoughts into media outlets. This culminated with Mitchell being put on trial in October 1925 for insubordination—namely, for stating that the crash of the airship USS *Shenandoah* was in fact the result of "incompetency, criminal negligence and almost treasonable administration of the national defense by the Navy and War Departments." Although the preceding is oft quoted, these were not the worst of Mitchell's remarks to the media. Mitchell further stated, "The bodies of former companions and buddies molder under the soil in America, Asia, Europe and Africa, many, yes many sent there directly by official stupidity. We all make mistakes, but the criminal mistakes made by armies and navies whenever they have been allowed to handle aeronautics show their incompetence."[17] Mitchell was unequivocally stating that current policies by Army and Navy officers caused the deaths of American airmen. Clearly, these statements could not go unchallenged by Mitchell's superior officers. He had crossed a line.

That being said, Mitchell knew exactly what he was doing when he called the reporters into his office that October morning. His were not the words of a grieving man or an outburst of emotion that happened to be caught by reporters. Rather, Mitchell's statement was an intentionally crafted response he spent hours refining. Mitchell knew what he was saying. The firebrand was pouring gas on a smoldering fire.

Mitchell's court-martial gained national attention in something akin to the O. J. Simpson trial of the 1990s. Spectators lined up by the hundreds, hoping to gain admittance to the courtroom. Here was an American general (colonel in actuality, although many of his supporters continued to call him general, his brevet rank during the war) being put on trial by the Army establishment. Mitchell's (second) wife joined him at the trial, which lasted seven weeks and concluded with Mitchell being stripped of rank, command, and pay and benefits for five years. In handing down the sentence, the courts-martial board wrote: "The Court is thus lenient because of the military record of the Accused during the World War." Only one of the judges voted to acquit Mitchell, Douglas MacArthur. Of his vote to acquit, MacArthur later said, "It is part of my military philosophy that a senior officer should not be silenced for being at variance with his superiors in rank and with accepted doctrine. I have always felt that the country's interest was paramount, and that when a ranking officer, out of purely patriotic motives, risked his own personal future in such opposition, he should not be summarily suppressed. . . . The individual may be martyred, but his thoughts live on." In this part of his memoirs, MacArthur could just as easily been referring to himself rather than to Mitchell. Rather than accept the punishment, Mitchell chose to resign from the Army.[18]

Carl Spaatz, who was present at the hearing, later remembered, "All of us who were friends of his—admired him—felt that the court martial was justified. He brought it on himself and I think he did it deliberately." Spaatz was correct. Mitchell wanted to use the trial as a way to create publicity for his forthcoming book, *Winged Defense*. Also present at the trial was Hap Arnold, who later recalled, "I can still remember how we all crowded into that room, with our wives listening while Billy's expert testimony turned out to be not the brilliant defiance we had looked for, but a dry reading to the committee of his own book *Winged Defense*." Arnold also echoed Spaatz's sentiments when he said, "Billy was licked, of course, from the beginning. . . . The thing for which Mitchell was really being tried he was guilty of, and except for Billy, everybody knew it, and knew what it meant."[19]

Mitchell's trial was important for another reason. While it highlighted the differences between ground officers and air officers, it created in the air arm a sense and acceptance of rogue leadership. From the moment Mitchell chose to resign, a cadre of junior officers aligned themselves with Mitchell and his ideals. Mitchell's acolytes now passed his ideas down through generations of officers, and Mitchell truly became both a martyr and a venerated figure in the future Air Force. The truth is the court-martial was a distraction from pushing independence and airpower doctrine forward.

Although to say it is still heresy to some in America's Air Force, the importance of William Mitchell to Air Force independence has been both seriously overstated and overinflated. Was Mitchell a singularly important figure in the history of airpower development? Obviously. Still, Mitchell was no icon. His opinions on the importance of any particular aspect of airpower seem to have stemmed from whichever posting he held at any given time, and they were expressed primarily to promote his own ideas and works. The Air Corps needed to move forward after Mitchell, but the die was cast. The martyrdom of Billy Mitchell gave rise to the zealotry of "air-mindedness" that would be pursued with religious fervor. With Mitchell now retired and permanently exiled from the military, the US Army Air Service continued to push the bounds not only of the possible—but in one memorable case, of the seemingly prosaic. Though mundane, the assignment ended in disaster.

The Air Mail Scandal and the Army Air Corps

Historians and journalists gave it many names: disaster, scandal, fiasco. Airpower historian Kenneth Werrell called it "probably the most important event in the Air Corps history between the wars." Chief of the Air Corps Benjamin Foulois cited the beginning date of his involvement in the operation,

9 February 1934, as one of the three "most significant dates in the development of American air power." Yet, this event remains virtually ignored despite the gravitas given it by airpower leaders at the time.

In 1934, President Franklin D. Roosevelt, responding to a Senate investigation into the improper awarding of contracts for mail delivery to major airlines, ordered the US Army Air Corps to begin delivering the United States post. To say the Army Air Corps was ill-prepared to take on this particular mission would be a vast understatement. Secretary of War George Dern, without consulting Army Chief of Staff Gen. Douglas MacArthur or Chief of the Air Corps Benjamin Foulois, assured the president that the Air Corps was up to the task of mail delivery, which from the perspective of both Roosevelt and Dern seemed like a relatively simple task.[20]

When Harlee Branch, an assistant postmaster general, called Foulois to his office that afternoon in February, the general assured him: "If you want us to carry the mail, we'll do it." At that exact moment, and without Foulois's knowledge, Postmaster General James Farley and Attorney General Homer Cummings were briefing Roosevelt on the proposed plan. Before Foulois had the opportunity to return to his office and brief MacArthur or Drum, Roosevelt issued an executive order ordering the War Department to take over mail delivery operations. MacArthur got to Foulois first and exploded, demanding to know why a reporter was asking him about this new executive order. Foulois was thunderstruck. MacArthur told him, "Foulois, this is all academic now.... You're on your own now, Foulois. Yell when you need help from me—and keep me informed. It's your ball game." It was not combat, nor could it aptly fit any descriptive operation previously conducted by the air arm, save for airlift, which had still not been developed into a major mission type, but the US Army Air Corps found itself hauling the mail.[21]

Foulois, given a fait accompli by the president, and authorization by MacArthur, set about creating from scratch a nationwide Army Air Corps mail delivery system. He divided America into three separate zones (eastern, central, western), each under the command of an Air Corps officer: Maj. B. Q. Jones, Lt. Col. Horace Hickam, and Lt. Col. Henry "Hap" Arnold.[22]

Air Corps pilots lacked night-flying experience, instrument experience, and experience flying through weather—a deadly triumvirate that eventually led to sixty-six accidents and thirteen deaths of Air Corps officers while completing only 65 percent of their scheduled deliveries. Air Corps pilots were also flying aircraft not specifically designed for cargo operations. Both pursuit and bomber aircraft were significantly retrofitted for the airmail operations, such as by welding shut bomb bays and expanding areas in already cramped pursuit planes to carry the actual mail. The final report of the eastern mail zone operations stated that the "suddenness of the emergency forced actual operations to be initiated before the practical technique of those operations could be determined." In other words, the airmail delivery operations were begun in too hasty a manner. This was but one critique among thousands found in the exhaustive 439-page report.[23]

The Air Mail scandal proved disastrous in terms of lives lost as well as reputation damage for the Army Air Corps. Eventually, it even cost Benny Foulois his job, as he was forced into retirement on 21 December 1935. Still, the entire enterprise, even the attempt to deliver the mail by air, highlighted the willingness of airmen to accept risk, conduct a difficult mission, and face challenges to prove the efficacy of the cultural icon that was the airplane. Perhaps most important, these airmen were not afraid to fail—and even sacrifice their lives— to prove to the American people the capabilities of airpower. Participants in the airmail operations included Elwood Quesada, Harold L. George, Laurence Kuter, and Ira Eaker, each of whom were destined to play even greater roles in the future of American airpower operations.

In the wake of the Air Mail scandal, the US Army convened another board. It was composed of the same five generals of the Drum Board, but this time Charles Lindbergh and Orville Wright were asked to join as well. Both refused, after which former secretary of war Newton Baker became the chair of the "Baker Board." In addition to Baker, additional members included Jimmy Doolittle (then manager of Shell Petroleum's aviation department) and Edgar Gorrell. The board came to largely the same conclusions that the Drum Board had and did not recommend any radical changes in Air Corps organization following the Air Mail fiasco. One notable exception was that Doolittle filed a separate minority report calling for the Air Corps to be "completely separated from the Army and developed as an entirely separate arm." He added that "if complete separation is not the desire of the committee, I recommend an air force as a part of the Army but with a separate budget, a separate promotion list, and removed from control of the General Staff."[24]

Just as the Air Mail scandal ended, a new cadre of younger officers, those graduating from West Point in the 1920s, found themselves transferred to Maxwell Field to attend the Air Corps Tactical School. This new generation of officers began attending the school at a fortuitous time as the first true theories of airpower were being read and discussed. They would make up the core group of men who would plan the air war for World War II, lead it through 1941–45, and then take on increasingly larger roles in a new independent Air Force into the early Cold War. Of course, none of them had any intimation of any of this. At that moment, they were primarily concerned with their posting to Maxwell Field, in the Deep South.

ACTS

As previously mentioned, in 1920, the US Army established a separate school for instruction of officers in aerial strategy and tactics. This school was the US Army Air Service Tactical School. After 1926, the Air Corps Tactical School, more commonly called ACTS—located initially at Langley Field, Virginia (1920–31) and later at Maxwell Field, Alabama (1931–42)—was the birthplace of all major air force doctrines brought to fruition in the Second World War. Conceived of as a preparatory school for the Command and General Staff School (CGSS) at Fort Leavenworth, its graduates hoped to move on to the latter school for promotion potential. Many found themselves disappointed when not chosen to move to CGSS at the completion of their year of aerial study. In 1929, ACTS took on a new motto: *Proficimus More Irretenti*. This translates to "We make progress unhindered by custom." No better motto could represent the school or its relationship with the larger US Army. The modern incarnation of ACTS, the Air Command and Staff College, particularly its Department of Airpower, continues to use this motto.[25]

The Personalities

Those who attended ACTS were a veritable "who's who" of the Army Air Forces and the founding generation of an independent American Air Force. The graduates included Carl Spaatz (1925); George C. Kenney (1926); Frank Andrews, George Brett, and Robert Olds (1928); George Stratemeyer (1930); Claire Chennault (1931); and Harold L. George (1932). Joining this group of notables was the next generation of airpower advocates. Younger by as much as a decade than their contemporaries, these men comprised a youthful vanguard that would become general officers once the Second World War broke out. This group included Muir S. Fairchild, Haywood Hansell, Laurence Kuter, and Hoyt Vandenberg (1935); Ira Eaker and Elwood Quesada (1936); and Lauris Norstad and William Tunner (1940). Of this august list, several stayed on as faculty members after graduation, including Chennault, Kuter, and Hansell. Air Force Historian Richard Hallion stated that the school was "the birthplace and nurturing ground for American air doctrine." ACTS in the late 1930s became the fertile ground in which bombardment theory grew. Although ACTS faculty taught courses on other types of aviation, bombardment became not only the dominant theory; it became the only theory that mattered.[26]

The Army's air arm in the interwar years did several things that not only paved the way for operations in the Second World War, but also set the stage for independence after the war. The Air Corps Tactical School taught airmen how to think about air warfare and inculcated in them a deep appreciation of bombardment theory and knowledge of how to turn that theory into operational plans. Although bombardment became the dominant theory, some airmen, such as Chennault and Quesada, remained committed to the pursuit school.

The move of ACTS from Langley to Maxwell demonstrated a need for the US Army Air Corps to separate itself from Washington, DC. Historian Craig Morris recognized the move to Maxwell Field provided "an out-of-the-way venue where students could focus on academic work without the interruptions that came with an operational flying base that was relatively close to Washington"—that is, Langley Field. The separation from the general vicinity of the nation's capital, and thus from senior military leadership, led to the impression that survived for decades after the move that Maxwell was a "sleepy hollow." In the 1920s and 1930s, nothing could be further from the truth.[27]

The ACTS faculty meticulously planned each one of their daily lessons. Instructors typed out their lessons nearly verbatim. Beyond the obvious theories, air war doctrine was covered. The overall course was divided into the Department of Ground Tactics—this was after all an Army course—and the Department of Air Tactics and Strategy. The latter department took up more than half of the year's course load and was divided in sections, which included the various types of Air Corps missions: Observation Aviation, Attack Aviation, Pursuit Aviation, and Bombardment Aviation. Classes ranged from the more theoretical "An Inquiry into the Subject of War," "The Objective of Air Warfare," "The Principles of War," and "The Aim of War" to the more tactical "Driving Home the Bombardment Attack," "Tactical Offense and Tactical Defense," and "Practical Bombing Probabilities."[28]

Some ACTS classes even had required classes in horsemanship. Field Artillery officers, used to working closely with the animals, like Larry Kuter enjoyed those afternoons, while many other officers, including Muir S. Fairchild, did not. The obvious fear, ironically enough, was that riding a horse was inherently dangerous and one might fall, suffer an injury, and thus be removed from the ability to fly an aircraft, itself an inherently dangerous profession.

Technological Development

Although much of this chapter has focused on personalities, the interwar Air Corps also saw vast technological advancements. In the time span of two decades, the Air Corps went from canvas biplanes to all-metal, enclosed-cockpit aircraft with such innovative features as retractable landing gear.

It was in a 1932 speech to Parliament titled "A Fear for the Future" that Stanley Baldwin spoke the famous phrase "The bomber will always get through,"[29] which became something of a mantra for ACTS faculty. Although remembered for that sole line, Baldwin's fuller speech gives a horrific picture of the future of aerial warfare:

> I think it is well also for the man in the street to realize that there is no power on earth that can protect him from being bombed. Whatever people may tell him, the bomber will always get through. . . . The only defence is in offence, which means that you have to kill more women and children more quickly than the enemy if you want to save yourselves. . . . If the conscience of the young men should ever come to feel, with regard to this one instrument that it is evil and should go, the thing will be done; but if they do not feel like that—well, as I say, the future is in their hands. But when the next war comes, and European civilization is wiped out, as it will be, and by no force more than that force, then do not let them lay blame on the old men. Let them remember that they, principally, or they alone, are responsible for the terrors that have fallen upon the earth.[30]

As he closed his speech, Parliament erupted in cheers. What Baldwin advocated for, and at the same time warned against, was the development of a fleet of bombers capable of bringing destruction to any point on the globe. Baldwin called the potentialities of airpower "incalculable and inconceivable."[31]

The bombardment proponents, the so-called bomber mafia, at Maxwell and those lobbying for the development of new bomber aircraft in Washington set out to make the incalculable and inconceivable a reality. Both bombers and pursuit aircraft developed in the prewar Air Corps during this period came in all manner of shapes.

The YB-9 produced by Boeing took flight in 1931. The monoplane bomber looked like a cigar with wings. It carried slightly over two thousand pounds of munitions. Although it was the first all-metal bomber designed for the Army Air Corps, the crew still flew in an open-air cockpit. One of the early fliers, Lt. Larry Kuter, remembered the plane's fuselage warped under pressure. If the pilot engaged one rudder while simultaneously moving the stick in the opposite direction, the torque created bent the aircraft. A more modern design followed with the B-10, and for the first time, the Army Air Corps had a bomber capable of pulling away from current pursuit models. Possessing a similar payload capacity to the YB-9, the B-10 at least enclosed the crew inside the aircraft. The B-18 Bolo was the first prewar bomber to see extensive service during the war, although primarily not in the role it was designed for. The bombardment proponents laid their faith at the altar of the bombers developed in the mid-1930s and that entered service at the same time: The B-17 and B-24, which are discussed in the following chapter.

For fighter aircraft, still being called pursuit in the interwar period, the Army Air Corps purchased the P-26 Peashooter. Today the Peashooter looks more like an exhibition plane than an operational military aircraft, but it remained in service as late as 1941. Following this was the Seversky P-35, the first all-metal, closed-cockpit fighter with retractable landing gear. The aircraft corporation's owner, Alexander de Seversky, eventually lent his pen to the development of airpower thinking, in the process giving America the phrase "Victory through Air Power" and a book so named, which inspired a Walt Disney film. Following the P-35 came the Curtis P-36. All of these aircraft, bombers, and pursuit fighters proved to be technologically outdated and easily outmatched at the beginning of the war.[32]

The Coming War and AWPD-1

ACTS not only adequately prepared airmen for the coming war, it instilled within them a deep respect for what airpower could accomplish, if only given the opportunity. Some became deeply entrenched doctrinaires whose adherence to strategic bombing would grow deep roots. Others focused on tactical employment of aircraft or on attack aviation. But no matter where they fell on the spectrum of how airpower could be employed—and there were zealots on all sides—they all agreed on what airpower could accomplish. They were "over not through" adherents. They earnestly believed that airpower could shorten the next war. Since the early 1920s, the courses at Langley Field, and later at Maxwell, brought together the writings of Douhet and Mitchell, but combined them with scientific, mathematical, and organizational principles that helped create a functioning air arm capable of use in war.

The opportunity arose in the summer of 1941 when several ACTS instructors found themselves assigned to Washington, DC, to serve Army Chief of Staff Gen. George Marshall and the head of the Air Corps, Gen. Hap Arnold. This group consisted of Laurence Kuter, Haywood Hansell, and Kenneth Walker. These four, under the direction of Lt. Col. Harold George and at the behest of Arnold, sequestered themselves away in the War Department, and in a scant nine days they created the first true air plan for use against an enemy of the United States. It represented the birth of the American air arm as an organization capable of independent action.

The air plan for World War II was developed as almost an afterthought when considered in combination with the existing Rainbow Plans (contingencies for a two-ocean war). In short, Generals Marshall and Arnold recognized the need for an air-centric munitions and basing annex to the already created and approved plans. It fell to George, Kuter, Walker, and Hansell to create the plan in a matter of days. Just as quickly as they produced a plan, the four then had to brief it, route it through the War Department, and get final approval for the document all the way up to President Roosevelt in the White House so that it could be added into the existing "Victory Plan." What they created was Air War Plans Division Document—plan number one, commonly called AWPD-1. It codified everything these four men had learned and later taught at ACTS. It was classroom teaching transformed into official doctrine. Historian David Johnson called the document the "quintessential expression of American strategic bombing theory." Kuter called the work on development of AWPD-1 "the sharpest advance ever experienced in the development of the American air power concept."[33]

The important aspect to remember here is that these four individuals used concepts learned at ACTS, based on experiences and, more importantly, beliefs that developed after the First World War. Here was the opportunity to use airpower to shorten the coming war and to save the lives of American soldiers on the ground. AWPD-1 was the first true "air" way of war developed and executed by airmen. Kuter remembered, "We concluded quickly that it would cost far more in blood, time and money to overcome that will by amphibious assault and land force invasion than by strategic air attack on their will and their ability to resist." However, the planners also knew that winning the war through the air alone might not succeed. Kuter added, "At the same time we recognized that we would have to structure our Air Force to provide all manner of support to Ground Forces if the Nazi-Fascists could not be defeated by air alone." Being pragmatic, the planners conceived of a strategic air force, but one capable of providing tactical support to ground forces.[34]

Here were two sides of the same coin. The AWPD-1 staff fervently believed in mid-1941 that it was entirely possible for an independent air arm conducting strategic bombing missions to win the war in Europe all on its own. Barring that, though, they needed to plan for an air arm capable of providing support to troops on the ground as they took the overland route to the heart of the Nazi Empire. Therefore, AWPD-1 needed to create both a strategic air force capable of conducting attacks against Germany's—and later Japan's—heartland and industrial base (using long-range strategic bombers) and a tactical air force capable of gaining air superiority and providing support to those on the ground (using pursuit aircraft, light and medium bombers, and air-to-ground attack aircraft). As far the four authors were concerned, the United Sates was capable of building and the Army Air Corps capable of employing both types of air forces.

AWPD-1 focused on five "divisions," more aptly tasks, that airpower advocates hoped their nascent fighting force could accomplish in the coming war: (1) to conduct air operations in defense of the Western Hemisphere, (2) to prosecute an air war against Germany as soon as possible, (3) to provide for strategic defense in the Pacific theater, (4) to provide air support for the eventual invasion of the European continent, and (5) after victory against Germany was secured, to concentrate maximum firepower against Japan.

AWPD-1 hoped to achieve all of the above and, on a more esoteric level, considerably more. Codesigner of the air plan Haywood Hansell later stated, "Proper selection of vital targets in the industrial/economic/social structure of a modern industrialized nation, and their subsequent destruction by air attacks, can lead to fatal weakening of an industrialized enemy nation and to victory through airpower."[35]

Shortly after approval, a copy of the plan ended up in print in the *Chicago Tribune* under the headline "F.D.R.'S WAR PLANS." For the price of two cents, people could essentially purchase their own copy of AWPD-1. The publication of the Victory Plan might have become a major scandal. The byline was dated 4 December 1941. The story became irrelevant seventy-two hours later. America was at war. The country had a plan for the air portion of this war. However, very little of what AWPD-1 called for actually existed. The "arsenal of democracy" was not yet at full power. When it did get up and running it produced thousands upon thousands of aircraft of every size and every type: fighters, bombers, cargo, trainers. It also created Armageddon.[36]

The legendary pioneers of the Air Corps—Spaatz, Doolittle, Eaker, and their leaders Foulois, Patrick, Fechet, and Arnold—all pushed the boundaries of what was possible to accomplish in the air. Even when one includes the second generation of Air Corps officers, those commissioned in the 1920s and 1930s, this still represented a relatively small population that could actually execute airpower operations. Although these officers participated in the defining moments of the early Air Corps—the "Question Mark" flight, the Air Mail scandal, and the court-martial of Mitchell—they were too few in number to form an air arm on a war footing.

When the war came, the United States Army Air Corps was going to need hundreds of thousands of men and women to climb into the cockpits of aircraft to be pilots, copilots, and navigators, to man machine guns, to become bombardiers, to maintain and fix the aircraft, and for an unknown number of these personnel to sacrifice their lives in the defense of American values. The sacrifices of American airmen in World War II transitioned the Air Corps from a continental organization incapable of delivering the mail into a global enterprise of fighters, bombers, attack aviation, and airlift. While the interwar period showed developments across air missions, doctrines, roles, and aircraft types, there can be little doubt that during the Second World War, the United States Army Air Forces were going to be led by men who believed in the war-winning strategy of strategic bombardment. Thus ends the early development of American airpower and begins the second air force way of war.

EPOCH II

Strategic Dominance, 1942–1975

They were called heavy or "strategic" bombers. The first ones were an earthy green, that unique military color known as "olive drab," or dusty tan. Later in the war, it was decided the olive drab paint was unnecessary; it weighed too much and served limited purpose. After that, they retained their shiny aluminum skin. No one would ever call them "sleek" or "streamlined," but they dwarfed most other aircraft. The first of them, the B-17 Flying Fortress, became the most recognizable symbol of American bombardment during the Second World War. It carried ten men and eight thousand pounds of bombs. The "arsenal of democracy" built over 12,000 of them. Its shorter, some might say uglier, counterpart, the B-24 Liberator, was known to be a more difficult aircraft to fly. In total, America produced nearly 20,000 of them. These two heavy bombers defined strategic bombardment for American forces in the Mediterranean and European theaters throughout the war. The aircraft that followed them would become iconic for a different reason.

They were mighty and massive. Eleven men fit, if not comfortably, then easily, inside. It was 99 feet front to back and 141 feet wingtip to wingtip. They glinted silver in the sunlight, and four massive "turbosupercharged" radial engines powered the behemoth aircraft into flight. In its internal weapons bay, it carried 20,000 pounds of high explosive and incendiary bombs or, if it was one of the specially equipped "Silverplate" varieties: one special-purpose bomb. The B-29 Superfortress became the primary heavy bomber in the Pacific theater of operations (PTO). The most famous of these, the *Enola Gay* and *Bockscar*, ushered atomic weapons into military operations, and the delivery of these weapons became the mainstay of United States Air Force planning, programming, and budgeting in Strategic Air Command for four decades.

Providing support to the mission of strategic bombardment were a myriad of fighters, medium bombers, light bombers, attack and transport aircraft; but in the eyes of senior Army Air Forces and later United States Air Force leaders, every other aircraft only supported the strategic bombers. Though many of the fighter pilots became legendary, there was no doubt in the minds of those who ran the Army Air Forces and later USAF that bombing the enemy was the mission of the organization. Gen. Curtis LeMay supposedly famously quipped, "Flying fighters is fun. Flying bombers is important."

Following the Superfortress came the B-36 Peacemaker, the B-47 Stratojet, and the B-52 Stratofortress. These were the United States Air Force's strategic bombers and were the backbone of America's nuclear force. Other aircraft, medium bombers and fighter aircraft designed to deliver their own nuclear payloads, did not fare as well, but one thing was clear during and after the Second World War: the air arm of the United States had a defined way of war: strategic bombardment. It became the raison d'être for an independent United States Air Force.

CHAPTER 3

World War II

What Hath Man Wrought?

There will be no distinction any longer between soldiers and civilians.

—*Giulio Douhet, The Command of the Air*

Writing in 1990, historian Caroline Ziemke said, "Strategic bombing is not mere doctrine to the USAF; it is its lifeblood and provides its entire raison d'être. Strategic bombing is as central to the identity of the Air Force as the New Testament is to the Catholic Church. Without the Gospels there would be no pope; without strategic bombing there would be no Air Force." The religious fervor with which airpower thinkers approached the Second World War meant that anything going against these teachings was considered heresy.[1]

The Air Force entered the Second World War with a clear idea of what the organization hoped to accomplish—the defeat of Nazi Germany and Imperial Japan through the use of strategic bombardment as codified in AWPD-1. This desire was undergirded with a clear and definable theory that came from years of rigorous study and development at the Air Corps Tactical School. That theory was this: high-altitude daylight precision bombing (HADPB) against industrial targets would degrade the war-making capability of the Axis powers and lead to a shorter and less bloody conflict. If conducted correctly, strategic bombing could hasten an end to the war or, it was hoped, bring about a decisive victory without the need for large-scale invasions.

Once the "arsenal of democracy" was fully functioning, the United States Army Air Forces found itself capable of executing numerous missions during World War II: not only strategic bombardment, but every other mission it was tasked to do—long-range pursuit, reconnaissance, attack, and material and personnel transport. Although strategic bombardment has certainly been written about the most, the Army Air Forces demonstrated a clear ability to conduct air superiority, close air support, interdiction, and attack missions throughout the theaters of combat. This was in no small part due to the massive mobilization in the United States. In fact, it may be the reason that, despite the "Germany First" pledge to the Allies, the United States never really had to play a holding game in the Pacific.

This chapter focuses on the actions of the United States Army Air Forces (USAAF) in each theater of the war—North African and Mediterranean, Southwest Pacific, European, and Pacific—and balances this against the massive war machine pouring out industrial equipment in the United States. In every theater, the Army Air Forces, while focused on strategic bombing, also ably provided air superiority, close air support, interdiction, and reconnaissance throughout the duration of the war.

In 1939 and throughout 1940, Army Chief of Staff George Marshall began to summon a new cadre of young and energetic military officers to Washington, DC. Included in the group were the AWPD-1 authors and former ACTS instructors Haywood Hansell and Larry Kuter. Inside the War Department, they joined Harold George and Kenneth Walker to form the nucleus of officers that planned the Army Air Forces requirements for the coming conflict. In short order, each of these midgrade officers found themselves wearing general's stars. The United States Army Air Corps also underwent revision. In June 1941, the Army Air Corps became the USAAF, one of three branches within the Army. Most importantly, General Arnold became chief of the USAAF. While not heading an "independent" military branch, Arnold now reported directly to Marshall and could operate America's "air force" independent of intervention by any Army officer other than Marshall. After the publication of AWPD-1 in the summer of 1941, and with America entering the global war in December 1941, Arnold sent this junior cadre of air planners out to get operational experience. Hansell and Kuter were dispatched to Europe, where America was beginning to build the 8th Air Force, and Walker was sent to the Pacific. There they would join Arnold's immediate subordinates: Spaatz in Europe and George Brett (followed by George C. Kenney) in the Pacific. The struggle to prove airpower's capabilities was joined in earnest.

Pearl Harbor and the Pacific

The American airmen stationed at Hickam Field and Wheeler Field on the morning of 7 December 1941 have to a certain and understandable degree been overshadowed when discussing the events of that morning. The iconic ships harbored at "Battleship Row" and the images taken from the attacking Japanese planes and other Americans on the ground seared "Pearl Harbor" into the minds of Americans. The billowing black smoke and wrecked superstructure of the USS *Arizona* and the explosion of the USS *Shaw* became among the most memorable images of the twentieth century, but the attack on the naval base was but one arm of a multipronged attack that morning.

At Wheeler, Hickam, and Bellows Fields, aircraft parked wingtip to wingtip in straight lines provided targets for the bombers, dive-bombers, and fighter aircraft that strafed the ramps, runways, and hangars. All branches suffered both aircraft and personnel losses that December morning, but the attacking forces failed strategically when they did not destroy either the aircraft carriers (which were not in port) or the fuel farms. Ships and aircraft were replaceable; buildings and runways, repairable. In the end, Pearl Harbor was a tactical victory, but a missed strategic opportunity for the Japanese.

In a feat of heroic airmanship, two pilots were able to get their aircraft off the ground while under fire and begin to engage the Japanese pilots. Kenneth M. Taylor and George Welch took off and began fighting the Japanese above Wheeler Army Airfield. The heroic actions of these lieutenants from the 47th Pursuit Squadron earned each the Distinguished Service Cross while providing an important morale boost to the nation after the devastating defeat. In a way, Taylor and Welch embodied the very maverickism of Billy Mitchell, as both took off and engaged the enemy without orders. Their heroism also represented the ideal of the fighter pilot: individual action against overwhelming odds. The ability and desire to *do something* was present on the morning of 7 December, but it was going to take time for the Army Air Forces to do much of anything, as Pearl Harbor was just the opening blow of an onslaught throughout the PTO.

Thousands of miles away across the vast Pacific, the situation in the Philippines was just as dire. Historian William H. Bartsch covers this in his history *December 8, 1941*. Despite MacArthur and his headquarters staff already knowing about the attack on Pearl Harbor, MacArthur made no concrete preparations or reprisal attacks. Both were mistakes. The leading airman in the theater at this time was Lewis Hyde Brereton. Brereton drove directly to MacArthur's headquarters, but MacArthur's chief of staff, Richard Sutherland, intercepted him before he could find MacArthur. Sutherland refused to let Brereton in to see the boss, and Brereton returned to his own headquarters without permission to launch either his fighters or his bombers against Japanese bases in the area. Two hours later, Brereton returned to MacArthur's headquarters but was again rebuffed by Sutherland, who informed him that no offensive was to be taken until the Japanese made the first offensive move—this hesitancy prevailed in spite of the fact that MacArthur and his staff knew of the Japanese attack on Pearl Harbor. Brereton furiously argued that the attack against American Forces in Hawaii was an offensive act, but to no avail. The matter of Brereton, Sutherland, and MacArthur's interactions and decisions remains a contentious issue more than eighty years later.[2]

The Japanese attack against American bases in the Pacific region was, by any standard, catastrophic. Beyond the loss of naval ships, the Japanese succeeded in destroying some two hundred aircraft at Pearl Harbor and Hickam Field alone. What occurred at Pearl Harbor paled in comparison to the hundreds of aircraft of the Far East Air Force destroyed in the Philippines on 8 December. The opening round went to the Japanese, but from a certain point of view, that proved to be their high-water mark. After Japan's victory at Pearl Harbor and the capture of British, American, and Australian bases into 1942, the tide rapidly turned against the Japanese. They believed the losses suffered by America's Pacific Fleet would set their enemy back by a half a year or more. This, too, proved to be a strategic failure on the part of Japan.

The Army Forces in 1942

The strategic situation for the Army Air Forces—nay, the entire Allied militaries—in early 1942 was bleak. Hitler held most of Europe. Japan was on the advance seemingly everywhere in the Pacific. Following the loss of American ships at Pearl Harbor, the British lost the HMS *Prince of Wales* and *Repulse* to Japanese aircraft. Wake Island and Guam fell in December 1941, and throughout the first months of 1942, Japan created a line of advance running from its home islands southeast to the Marshall and Gilbert Islands and southwest nearly to Australia. Japanese forces controlled the Philippines, the Dutch East Indies, and a portion of New Guinea. On 27 February, while transiting toward the defense of Java, America's first aircraft carrier, the USS *Langley*, was attacked by Japanese aircraft and disabled. The carrier was then torpedoed by its two escorting destroyers to keep it from falling into Japanese hands. It was the first loss of a US carrier in the war. More importantly, as the Japanese advanced, they destroyed the bulk of the US Army Air Forces aircraft meant to blunt their advance. As Japan advanced throughout the Pacific theater, American airmen made diligent, but largely futile stands against the Japanese onslaught, equipped as they were with aircraft inferior to everything the Imperial Japanese Army Air Service had. The fight was lopsided. Tactics for defeating the Japanese Zero had not yet been devised, and in early 1942 it simply outclassed everything the American air force had.[3]

Bartsch noted that although more modern aircraft were being funneled into the Pacific throughout 1941, including B-17 bombers and P-40 fighters, there still existed a significant number of aircraft ill-suited to the tasks at hand. These included B-18 Bolos and Seversky P-35s, both of which were at the end of their technological life cycle and military utility.[4]

Back in the nation's capital, the winter and spring of 1942 was a time of confusion and expansion. Ira Eaker departed Washington in February along with six officers to begin the setup of American bombing operations in Europe. Carl Spaatz followed him to England in May. March saw the realignment of the US Army into three new forces: Army Service Forces, Army Ground Forces, and Army Air Forces. This reorganization resulted in a significant amount of autonomy for the United States Army Air Forces within the Army. From a small core of senior officers, the US Army and US Army Air Forces had to expand at an exponential rate to meet the unknown challenges of a global war, but before that happened, the United States received a strategic, if tactically insignificant, shot in the arm.[5]

The Doolittle Raid

American reprisals came quick, much quicker than the Japanese leaders believed possible after their string of successes. Japanese military leaders hoped the attack on US forces throughout the Pacific region would set American response efforts back by six to eight months. This proved incorrect. The response came a scant four months later, when sixteen B-25 Mitchell bombers, under the command of Col. James "Jimmy" Doolittle, launched from the deck of the USS *Hornet* for an attack on the Japanese mainland. The plan called for the *Hornet* to approach the Japanese coast, launch the bombers, and make a speedy escape from hostile waters while the bombers flew on to strike industrial targets in several major cities. Following the successful bombing runs, the planes would then fly across Japan to China and recover at bases there. Early detection by the Japanese ship *Nittō Maru* forced a premature launch that significantly altered the bombing timeline. As a result, the planes would certainly not have enough fuel to land as planned at the bases in China.[6]

Doolittle chose to launch the attack anyway. The attacking force cast a wide net. While most bomber raids struck targets in a single city, the Doolittle Raid attacked five different cities: Tokyo, Yokohama, Nagoya, Kobe, and Osaka. In his autobiography, *I Could Never Be So Lucky Again*, Doolittle recalled, "The primary purpose of the raid we were about to launch was psychological. The Japanese people had been told they were invulnerable. Their leaders had told them Japan could never be invaded." Striking so many targets sent a clear message to the people of Japan, one that would be brought home with cataclysmic results later in the war: the Japanese homeland was not invulnerable. Doolittle added, "It was the kind of touché the Japanese military could understand."[7]

After hitting their targets, the B-25s headed in the general direction of Chuchow (Quzhou), China, but had to do so by dead reckoning as the radio beacon located there was turned off. Fifteen of the sixteen aircraft crashed or crash-landed in China, and most of the crews bailed out of their aircraft. Sixty-nine members of the raid, including Doolittle, eventually made their way back home. Of the others, three were killed in action and eight became prisoners of war. Operationally, the attack succeeded in causing only minor damage. Strategically and publicly, it sent a signal that the Japanese homeland was not immune to attack, and it provided a psychological boost for the American people. American airpower—along with naval assistance—had struck back at the Japanese aggressors. A mere sixteen airplanes struck back at Japan, but it proved to be a heavy propaganda victory for the United States and a psychological blow to Japan.

The "Arsenal of Democracy" and the Emergence of Women Air Service Pilots

No way of war in the air was possible without the production of aircraft. America's industrial might during the Second World War produced a grand total of 299,230 aircraft of all types—bombers (very heavy, heavy, medium, and light), fighters, reconnaissance, transports, and trainers—between 1940 and August 1945. Once the United States entered the war in December 1941, the total production of aircraft increased each year: to 47,675 in 1942, 85,433 in 1943, and 95,272 in 1944. Between October 1943 and October 1944 alone, more than 1,000 heavy bombers rolled off the assembly lines every month. In roughly the same time span, more than 3,000 fighters entered service every month. The biggest single monthly total was in December 1943, when more than 8,000 aircraft—or 258 aircraft every day—entered service.[8]

Aircraft coming off production lines then had to be transported to airfields near the coasts before being ferried again across the Atlantic and Pacific Oceans to their final bases in theater for use in combat. It is at this point in the history of America's air force that women pilots emerge within the air arm. Pilot allocation was becoming a problem, at least in terms of the allocation of available male pilots. There were, early in the war, simply not enough qualified pilots to fill all the necessary roles: combat, trainer, and ferrying. The Air Force turned to women pilots, as epitomized by Nancy Love and Jackie Cochran.

Recent scholarship, most notably by Sarah Byrn Rickman and Katherine Sharp Landdeck, has finally given the contributions of women pilots of the Women's Auxiliary Ferrying Squadron (WAFS), Women's Flying Training Detachment (WFTD), and Women Airforce Service Pilots (WASP) their just historical treatment. Although the name of Jackie Cochran remains relatively recognizable, other women pilots including Nancy Love have also risen to the historical forefront as the vanguard of gender-norm-breaking leaders of the women's air service. During the course of the Second World War, thirty-eight women in the WAFS and WASPS gave their lives in service to their country, but it took until 2009 for their sacrifices to be officially recognized at the national level when then-president Barack Obama bestowed on them the Presidential Medal of Freedom. A year later, the US Congress awarded them the Congressional Gold Medal.[9]

Historian Molly Merryman noted, "The WASPs directly questioned the purportedly natural and expected status of men within the military by serving in one of its most desired roles.... WASPs challenged assumptions of male supremacy in wartime culture." Although other nations, most notably the Soviet Union, had a somewhat more progressive view of women air force pilots, the role the WASPs played in creating a larger world for women aviators cannot be overlooked, despite the fact their service was largely ignored for years after. It is important also to note that, despite Soviet airwomen having distinguished themselves in combat, a major reason for their role was not only to address the USSR's pilot shortage but to gain propaganda value for communism. More contemporarily, several WASPs have been recognized by the Air Command and Staff College at its annual "Gathering of Eagles" for making significant contributions to "air, space and cyberspace power."[10]

Arguments arose and battles were fought in 1941–42 over the exact nature and role women pilots could fill. Eventually two separate organizations emerged: the WAFS (under Nancy Love) and the WFTD (under Cochran). Both groups came into existence in September 1942. Although much has been made of the "Love-Cochran rivalry," both wanted women to serve as commissioned officers in the USAAF. However, both were denied this as their pilots were considered part of the US civil service and not active-duty members of the military.[11]

Any direct confrontation between the two women effectively ended in August 1943 when Jackie Cochran became the director of the women pilots once the training program and Ferrying Division pilots were combined into the Women's Airforce Service Pilots, and Nancy Love became the director of all ferrying operations. Although the program was inactivated in 1944, it proved that women were more than capable of flying all types of aircraft. Twenty-five thousand women applied to become pilots, and over one thousand completed their training. Cochran noted in her autobiography that the WASPs did much more than move aircraft from point A to point B:

My training program, my operations, the WASPs—everything worked and in fact, it worked extremely well. We were proving a point about women fliers, providing competent pilots to the ferry operation, towing targets—a nasty job—working on tracking and searchlight operations, simulated strafing, smoke laying, and performing other chemical missions, radio control flying, taking care of basic and instrument instructing, engineering test flying, as well as handling the administrative and utility flying. We had to prove points as basic as the fact that women could fly during any time of the month. Menstrual cycles didn't upset anyone else's cycle.[12]

These women provided a service to their country, but in doing so engendered a degree of contempt from many in society. After the war, they returned to their lives with dignity and rarely spoke of their accomplishments as pilots. While the Soviet Union used women pilots for combat operations, none of the members of America's ferrying division, the WASPs, saw combat or were allowed to ferry aircraft into a combat theater.[13]

Preserving the Story

World War II also found the Army Air Forces taking an active step toward the preservation of the history occurring in and around the organization. History suddenly became a mission of the Air Force. This new interest stemmed from two sources. First, in early March 1942, President Roosevelt dictated in a letter to the director of the Bureau of the Budget: "I am very much interested in the steps that you have been taking to keep a current record of war administration. I suggest that you carry the program further by covering the field more intensively, drawing on whatever scholarly talent may be necessary."[14]

Second, Gen. Hap Arnold took an active interest in the preservation of Air Force history and the war record of his organization. On 23 June 1942, Arnold directed Maj. Gen. Millard F. Harmon, chief of the Air Staff, and Col. Edgar P. Sorenson, Harmon's assistant chief for A-2 (intelligence), to keep "a running account of Army Air Forces participation in all military actions in all theaters." The next month when Brig. Gen. Laurence Kuter, now deputy chief of the Air Staff, issued a directive to the director of Organizational Planning "that action be initiated immediately to establish within the Air Staff an historical section capable of writing an accurate military history of the Army Air Force. It is important that our history be recorded while it is hot and that personnel be selected and an agency set up for a clear historian's job without axe to grind or defense to prepare."[15]

Thus, the historical program of the Army Air Forces was launched just as the US war effort began in earnest over the European continent and in North Africa. These directives prompted the efforts at historical preservation that allowed the USAAF's historical program to publish the seven volumes of the *Army Air Forces in World War II*, the *Combat Chronology 1941–1945*, and *Combat Squadrons of the Air Force in World War II*. Kuter himself would become a part of this history when he received orders first to England and shortly thereafter to North Africa.[16]

Mediterranean and Middle East Theater

As much of the history of the war focused on strategic bombing and especially the Combined Bomber Offensive (CBO), it is easy to overlook the application of airpower in North Africa and the Mediterranean. In the summer and fall of 1942, as American airpower was being built up to conduct operations against targets on the European mainland, the first major operations for the USAAF occurred in North Africa as part of Operation Torch.

Over the protests of some USAAF leaders, aircraft were hived off the 8th Air Force to create the 12th Air Force, under the command of Jimmy Doolittle, for operations in North Africa. Even as the 8th was building up its forces in England, the 12th was doing the same. It is fitting that what follows is a discussion primarily of tactical aircraft and medium bombers, although the larger B-17s and B-24s also participated. This somewhat flies in the face of what many have written, including this author, with regard to interwar airpower theory. As historian Chris Rein noted in his book *The North African Air Campaign*, the campaign in North Africa "refutes the suggestion that the USAAF so focused on the strategic bombardment mission that it neglected tactical developments." Again, the USAAF was able to do both—strategic and tactical operations—simultaneously. Although airpower leaders often bristled at strategic bombers being used for tactical employment, there is no doubt that these aircraft were often successful in this role.[17]

The USAAF assets allotted for Operation Torch, as the American landings in Morocco and Algeria were called, seem paltry in retrospect: two heavy bombardment groups and four fighter groups. Even the official Air Forces history of the war calls the air support to the initial landings a "minor one," but the USAAF contributions expanded greatly as the campaign progressed. American airpower in North Africa began with a strike from the sea. Having no bases in North Africa from which to launch American aircraft, the first strikes of Operation Torch came from aircraft carriers, carried out primarily by US Navy and British aircraft. The USAAF placed American fighters and bombers of the 2nd Air Force at Gibraltar and at locations in England to fly into airfields as soon as they were captured and secured. The targets for the opening salvos were the landing ports, which sorties were to hit while also providing direct support to the attacking forces on the ground.[18]

Not all of the attacking forces came from the US Navy. In one of these lead flights, William Momyer, commander of the USAAF 33rd Fighter Group, took off from the carrier USS *Chenango* with seventy-seven P-40s. As he approached the runway at the Port Lyautey airfield, Momyer came under heavy fire and crash-landed. Momyer rapidly assessed the situation while under sniper and artillery fire and determined that no more of the 33rd could land at Lyautey that day. Only one other aircraft of Momyer's group attempted to land, and it also crashed. Momyer rescued his fellow pilot from his flipped and burning aircraft. Ironically, the reason no aircraft made a safe landing was that the runway was heavily damaged from previous naval bombardment. Such was USAAF's contribution to the landings, or as the official history stated, "Thus the US Navy's carrier aircraft had assumed the whole burden of air cooperation with the Western Task Force."[19]

The US Navy provided the bulk of the airpower in the first few days of Operation Torch with the intention of allowing American ground forces to occupy land, including airfields, that could then receive USAAF aircraft to continue the attacks. As soon as ground forces secured airfields, flights departed from Gibraltar to make North Africa their new home. From there, the USAAF pilots followed their ground brethren as the latter seized territory and airfields from the Vichy French and Axis forces. The Germans continued to push men and machines into North Africa in late 1942 and early 1943, but to no avail. American and Allied aircraft gained control of the air and held it, which made resupply missions into North Africa tenuous, and in some cases deadly, for the Germans. During Operation Flax in April 1943, Allied aircraft decimated German resupply transports off the Libyan coast. The deputy commander of the Northwest African Tactical Air Force, Brigadier General Kuter, noted, "By dark the Tunis harbor was littered with some floating air cargo, ditched aircraft and small rescue boats. On Palm Sunday our claims totaled over 100 transports and ten fighters." This included both air and sea transports. As historian Chris Rein noted, "Allied dominance of the skies, and therefore of the seas, prevented any Dunkirk-style mass evacuation, with the result that most of the Axis troops rushed into North Africa between November 1942 and May 1943 became casualties, either killed[,] wounded, or captured."[20]

Once US forces held a solid foothold on the continent of Africa, the North African campaign began in earnest. Carl Spaatz now reorganized his combined air arm into the Northwest African Air Forces (NAAF), composed of the following units: the aforementioned Northwest African Tactical Air Force (NATAF), the Northwest African Strategic Air Force (NASAF), the Northwest African Air Service Command (NAASC), as well as a reconnaissance arm, training arm, and the Northwest African Coastal Air Force (NACAF). It is important to note that the NAAF under Spaatz was only one of five separate air commands operating in the Mediterranean theater. Eventually these disparate forces were consolidated and brought under the Mediterranean Air Command commanded by senior RAF commander Sir Arthur Tedder. The USAAF made a good showing in North Africa, but much needed to be learned in the areas of cooperation between the Allies, and also between the services. Most of the command arrangements eventually became codified doctrinally in War Department Field Manual 100-20, "Command and Employment of Air Power." Understandably, ground commanders wanted to "own" the air force assets supporting their ground troops; however, this made airpower over the entire army inherently inflexible. "Tying" any particular air unit to a particular commander on the ground meant that it was incapable of responding elsewhere where it might be needed.[21]

FM 100-20 put a stop to this practice. Official doctrine of the United States Army now stated: "Land power and airpower are coequal and interdependent forces; neither is an auxiliary of the other." Furthermore, it stated that airpower would fall under direct control of the "superior commander" who held overall responsibility for all forces in a theater and that this superior commander "will not attach Army Air Forces to units of the ground forces under his command except when such ground force units are operating independently or are isolated by distance or lack of communication." FM 100-20 established doctrinal practices that remained in place throughout the war.[22]

With operations in North Africa progressing, senior leaders from the United States and the United Kingdom met at Casablanca to determine the next steps in the war. At the Casablanca Conference, held 14–24 January 1943, the Allies agreed not only to the invasion of Sicily, but also to the Combined Bomber Offensive. In a memorandum dated 21 January 1943 from the combined chiefs of staff (CCS) of both countries and addressed to "the appropriate British and US Air Force Commanders," the CBO directive called for "the progressive destruction and dislocation of the German military, industrial and economic system, and the undermining of the morale of the German people to a point where their capacity for armed resistance is fatally weakened." The directive laid out five target sets: German submarine construction yards, German aircraft industry, transportation, oil plants, and "other targets in enemy war industry."[23]

More importantly, the directive gave senior air leaders or overall theater commanders a wide degree of latitude in where to place their bombing emphasis: "The above order of priority may be varied from time to time according to developments in the strategical situation." In other words, each air commander had discretion to make his own decisions about what targets to attack and in what priority. This would obviously change over the course of the war. The CBO directive signaled out one city for attack: Berlin. The German capital "should be attacked when conditions are suitable for the attainment of especially valuable results unfavorable to the morale of the enemy or favorable to that of Russia." The Combined Bomber Offensive gave Hap Arnold exactly what he wanted: the 8th Air Force turned loose over Germany and German-occupied France. Between the CBO and FM 100-20, airpower and the US Army Air Forces became virtually independent. It at least had autonomy from most ground commanders, and with the exception of overall theater commanders, very few individuals could dictate what they wanted the USAAF to do. In the meantime, the Casablanca Conference also set the next step in the march toward the German homeland: the invasion of the island of Sicily. Here, airpower played a further important role in destroying the Luftwaffe and in striking targets on the ground.[24]

Historian Alexander Fitzgerald-Black argues in his book *Eagles over Husky* that Allied airpower in Sicily made a critical contribution to the wider narrative of the air war during the Second World War. Although he points out that some historians consider Sicily somewhat of an operational blunder because German forces were allowed to escape the island, Fitzgerald-Black argued that "Allied strategic success in Sicily and the Mediterranean in mid-1943 mattered far more than the failure to prevent German forces on the island from escaping."[25]

Operation Avalanche

Once the Allies secured Sicily, it was on to Europe and the invasion of Italy, although there continued to be debate about the efficacy of invasion of the continent from the south versus the planned invasion across the English Channel. Churchill had always favored an attack into the "soft underbelly" of the European continent, but the Sicily campaign proved this was clearly a misnomer. Even so, by late 1943 with North Africa and Sicily secured and the Mediterranean opened to allied sea power and shipping, and with the understanding that the invasion of France could not take place until the spring or summer of 1944, the Allied leaders recognized that doing nothing was not really an option. Thus, Operation Avalanche began in September 1943. Any German division engaged against the Allies in Italy was not a division capable of supporting the war in the East; thus an invasion of the peninsula would relieve the pressure on Soviet forces. With seventeen divisions in Italy and another fifty awaiting the Allied invasion of France, a substantial portion of the German army could not be brought to bear against the Soviets. The same was true of German airpower, which was now stretched to the absolute breaking point. The invasion of Italy began on 3 September with Allied forces traveling from ports of embarkation in both North Africa and Sicily to landing zones in Salerno, Calabria, and Taranto.

This was the opening stage of a campaign marching northward against a series of German defensive lines that would last well into the summer of 1944.[26]

The air war in Italy proceeded in much the same fashion as in North Africa. The USAAF supported the operation from bases in North Africa and Sicily and waited for Allied ground units to get ashore and capture airfields needed to continue the march up the Boot. In supporting the initial landings, air forces destroyed what was left of the Luftwaffe in Italy. According to historian Robert S. Ehlers in *The Mediterranean Air War*, "The air armada destroyed 903 Luftwaffe aircraft at a cost of 205 planes. It was a telling victory and the last time the Luftwaffe made any kind of appearance over the battlefield."[27]

With the capture in late September of Foggia, and the Axis air base located there, the Allies extended their reach deeper in Europe. The German military and the civilian population living near strategic and industrial targets now had to contend with Allied strategic airpower from two directions. The Mediterranean Allied Strategic Air Forces now joined the attack into the heart of Europe and expanded the CBO against Germany. Allied strategic airpower in Italy was significantly closer to the targets in southeastern Europe, which included strategic targets in the Balkans and the oil fields in Romania, notably at Ploesti.

The Luftwaffe made a final all-out effort against the American forces trapped on the beaches at Anzio in February 1944, but failed to dislodge it. Although the war in Italy continued to grind on for the remainder of the war, the German Luftwaffe had shot its bolt. James Corum, a historian of the Luftwaffe, noted, "For two years the Allies would be forced to fight their way up the Italian peninsula, taking heavy losses all the way." Allied airpower was often limited in its support to the ground armies due to the terrain and weather, but their opponent fared no better. Ehlers stated, "The Luftwaffe's campaign in Italy in 1943–1944 is a snapshot of a force in rapid decline."[28]

ETO and the CBO

In the course of the war, the combined efforts of the Royal Air Force and United States Army Air Forces used over 28,000 bombers and launched more than 1.4 million bomber sorties against targets throughout Europe. Later in the war, an additional 2.6 million fighter sorties added to these numbers. The bombing campaign dropped more than 2.7 million tons of bombs and destroyed 3.6 million buildings. The USAAF and RAF lost an equal number of airmen in these raids, roughly 79,000 each. Over 300,000 civilians died in these attacks and another 7.5 million were left homeless.[29]

Americans began bombing the European continent on 17 August 1942 when twenty-three B-17s led by Col. Frank A. Armstrong in the B-17 *Butcher Shop*. Armstrong's copilot was Maj. Paul Tibbets, who was destined to become the most famous bomber pilot of the Second World War. Ira Eaker flew as an observer on the B-17 *Yankee Doodle*. The target was a railway marshaling yard at Rouen, in northern France. Antiaircraft fire was light, only one B-17 was damaged, and the bombing itself, especially compared to later missions, was surprisingly on target. The strategic bombers of the United States Army Air Forces had entered the war. But what did this singular mission portend for the future of the war in Europe? The short answer is not very much. The early and difficult efforts of the American bomber force in 1942–43 were later famously turned into the film *Twelve O'Clock High* which begins with the early bombing raids and culminates in the attacks of "Black Thursday" against the ball bearing plants at Schweinfurt. Not only would this film be later added to the National Film Registry of the Library of Congress, but it became a mainstay of US Air Force officer education programs.[30]

Craven and Cate's *Army Air Forces in World War II* notes, "When the small force of B-17s from XII Bomber Command took to the air on 17 August 1942, they carried with them much more than a bomb load of trouble for the enemy. They carried with them a long heritage of debate and controversy." The specter of protracted trench warfare compelled Army Air Corps members to reach for an "over and not through" solution after World War I. It is unlikely any of them ever considered that what they created was, in some respects, even worse.[31]

The strongest adherents to this "over not through" mentality were typified by the likes of Alexander de Seversky, the aircraft designer turned author who held the enmity of Hap Arnold. De Seversky offered what historian Michael Sherry called "an almost effortless path to victory." This path was simple enough, and it was espoused by instructors at the Air Corps Tactical School. Bombing Germany's (and Japan's) war-making industry, they had claimed, would cause economic collapse and bring Germany to its knees.[32] The previously mentioned CBO directive of 21 January 1943 effectively put ACTS doctrine into an order.

Spaatz stated years later, "We always attacked only legitimate military targets with one exception—the capital of the hostile nation. Berlin was the administrative and communications center of Germany and therefore became a military target."[33]

The Pointblank directive, issued on 10 June 1943, modified the CBO somewhat. It stated: "Since the issue of this [CBO] directive there have been rapid developments in the strategical situation which have demanded a revision of the priorities originally laid down"—by which it alluded to the massive shift underway by German defensive fighters, both day and night, to intercept and destroy the incoming bomber streams. This opposition had developed to the point, the directive predicted, that "unless this increase in fighter strength is checked we may find our bomber forces unable to fulfill the tasks allotted to them." Airpower proponents now acknowledged the cruel reality that went contrary to all the teachings prior to the war. Unless Allied fighter aircraft stopped the Luftwaffe from attacking the bombers, the bombers were not going to get through.[34]

In accordance with that realization, the directive ordered the 8th Air Force to modify the CBO and attack the "intermediate objective" of German fighter strength by going after "German airframe, engine and component factories and the ball-bearing industry of which the strength of the German fighter force depend." This did not work, however. German aircraft production continued to increase throughout the war. The strategic bombardment of these target sets was not achieving its goal. What did work was the release of Army Air Forces fighter aircraft to go after the German fighters in the air and on the ground. The pragmatic decision to adopt this approach, which proved effective, was made by legendary airman Jimmy Doolittle once he took charge of the 8th Fighter Command.[35]

The Pointblank directive noted: "American fighter forces will be employed in accordance with the instructions of the Commanding General, Eighth Air Force in furtherance of the bomber offensive and in co-operation with the forces of Fighter Command."[36]

In December 1943, the P-51 Mustang escorted US bombers for the first time. The inclusion of drop tanks on the pursuit aircraft allowed them to fly farther and engage with the enemy deep in his own territory. Among the bomber proponents teaching at ACTS prior to the war, who had latched on to the mantra that the bomber would always get through, there were regrets at their shortsightedness. Gen. Larry Kuter admitted much later in life that "there were regrets. No one regretted it more than the bombers." With the addition of drop tanks on the fighter aircraft, the war took a drastic turn: the ability to attack the Luftwaffe entered a new stage, but this would require changes to the current doctrine that kept the fighter aircraft welded to the bomber formations. That would have to come later.[37]

One of the biggest surprises to air leaders of the Second World War was the heavy loss of the bombardment crews. The previously conceived idea of saving lives on the ground by shortening the war through bombardment of the enemy homeland resulted in the wholesale slaughter of Army Air Forces aircrews. In the 8th Air Force, 26,000 men lost their lives as part of the more than 47,000 casualties accrued during the war. These men fell to their deaths in one of the 10,561 aircraft lost over the skies of Europe between 1942 and 1945; these numbers do not include C-47s lost as part of combat operations.[38]

The losses on the ground proved even more horrific as more than 300,000 died and an additional 780,000 wounded.[39] These were not all ardent Nazis. These were not the SS. These were husbands, wives, daughters, and sons who happened to live in the towns that contained industrial, military, and economic targets. All were tied to the Nazi regime no matter where their allegiances lay. Total war arrived via American bombers.

The War in the Air

During the course of the Second World War, the USAAF lost 17,275 bombers. For every B-17 lost, ten crewmen went down with it. For every B-24 lost, eleven crewmen were lost with it. If they were lucky, the crewman could parachute out, almost always over enemy territory, to spend the rest of the war in captivity or, if they were lucky, escape back to England. The 8th Air Force suffered 47,000 casualties during the war, of those 26,000 were killed in action. Life in the sky for members of a bomber crew was nothing short of horrific and terrifying. Thousands of feet in the air, there was absolutely nothing a crewmember could do to protect himself. While feelings of self-preservation and support to other crewmembers clashed inside, there was little a man could actually do other than his assigned task. If no enemy fighters showed themselves, especially over the target cities, the gunners were simply along for the ride with little else to do than watch the flak as it leaped out to destroy their aircraft.

The air war changed over time, and to the proponents of strategic bombardment it seemed they never had enough time to prove themselves. An ever-changing strategic and operational situation continually forced heavy bombers to be pulled away from the targets the bomber leaders wanted to attack. Instead of prolonged, concentrated attacks against the industrial centers, the heavy bombers were routinely sent against other sites: submarine pens, transportation targets, and direct support to invasion forces—all of which the "bomber mafia" viewed as a misuse of strategic airpower. In the minds of men like Arnold, every strike against a tactical target was a strike not made against a strategic one, and according to their doctrine, this only lengthened the war.

The CBO also highlighted and reinforced the differences in British and American approaches to strategic bombing. The American bomber forces continued to use their bombers in ever-growing numbers in the prescribed high-altitude daylight precision bombing taught at ACTS. The British continued with their preference for nighttime area bombing. The CBO also demonstrated both the potential and the limits of airpower.

1942

American operations against targets in Europe began during the summer of 1942 and ramped up through the fall and winter. The buildup of American strategic bomber fleet proved to be a slow process, not only production of the aircraft themselves but also the training necessary for their employment. Senior

Army Air Forces officers of the 8th Air Force found themselves chagrined that as soon as some aircraft arrived, the planes were redirected to the 12th Air Force in preparation for Operation Torch.[40]

The first wave of the 8th Air Force headed for England on 27 April 1942 not in aircraft, but aboard the transport ship *Andes*. Approximately 1,800 men of the headquarters and other commands sailed across the Atlantic, while the bombers and pursuit aircraft were being prepared to depart the United States in June.[41]

Only eighteen B-17s flew the first raid against Rouen on 17 August. It is important to note that even a raid of this size placed 180 American airmen in harm's way. Even the smallest raids against German-occupied France placed hundreds of men in danger, not to mention the fighter escorts. As previously mentioned, the raid against Rouen was successful in that all the bombers returned and battle damage was minimal.[42]

1943

By December 1942, and into the early months of 1943, American innovation became readily apparent. Brig. Gen. Laurence Kuter, on his way to an assignment in North Africa, told his replacement at the 1st Bombardment Wing, Brig. Gen. Haywood Hansell, that one of Kuter's (now Hansell's) group commanders, Col. Curtis LeMay, had developed new formation tactics that led to better formation flying and bombing accuracy. This new "box" configuration rapidly became the prescribed formation for all bombing missions. LeMay's innovation, coupled with changes still to come for American escort fighters, altered the air war over Europe.[43]

With the launch of the CBO and the influx of aircraft and personnel, the raids increased in size and intensity, and became significantly deadlier for all concerned—not just for the industrial targets on the ground, but also for the workers inside those targets, the American and British airmen in the skies above, and the civilians living near the targets.

Perhaps the most ignominious bombing mission was Operation Tidal Wave against Ploesti in August 1943. Less than a year after the first raid against the Rouen marshaling yards, the USAAF launched a raid composed of 177 B-24s against the oil refinery near the Romanian city. Of those aircraft launched against the target, 53 would not return. Eschewing the high-altitude method, the B-24s came in low. Rather than being knocked permanently out of action, the refinery was back up and running a month later. A hard lesson was learned. Hitting a target and temporarily disabling it was one thing. Keeping it out of action required successive bombing missions. The concept that a single refinery, or any target for that matter, could be bombed once and thereby removed as a war-making asset for the enemy proved to be farcical. It was necessary to hit a target many times over to keep it out of action. The buildup and expansion of the 8th Air Force thus continued throughout 1943.[44]

As the official history of the Army Air Forces in World War II noted, "The Eighth Air Force, whatever the imperfections still existing in its organization, enjoyed the benefit of more than a year of hard-won experience in the theater and required chiefly the necessary men and planes to prove itself an effective instrument of strategic bombing." With the Pointblank directive in effect and an ever-increasing number of bombers of various sizes and their associated crews continuing to be built up in England, the USAAF was in a position to begin making the decisive difference so long dreamed about by the advocates of strategic bombing.[45]

1944

In early 1944 as the Allies prepared for the long-awaited invasion of France, the strategic picture in the air changed significantly. In January, Doolittle took command of the 8th Air Force from Ira Eaker and ordered his fighter aircraft to go after the Luftwaffe. When Doolittle walked into the office of 8th Fighter Command, he saw a sign stating, "THE FIRST DUTY OF THE EIGHTH AIR FORCE FIGHTERS IS TO BRING THE BOMBERS BACK ALIVE." In other words, existing tactics needlessly tied fighters flying escort missions to the bomber streams. Doolittle ordered the sign taken down and replaced with one that read, "THE FIRST DUTY OF THE EIGHTH AIR FORCE FIGHTERS IS TO DESTROY GERMAN FIGHTERS." Doolittle loosed his fighters against those of the Luftwaffe. With the aid of drop tanks and the freedom to go on the offensive, the fighters changed the status quo in the Allies' favor in the air war.[46]

This change coincided with Operation Argument, more commonly remembered as "Big Week," which began on 22 February 1944. This was a dual attack: against German aircraft factories, such as manufacturing and assembly plants, on the ground and the Luftwaffe fighters that rose to join the battle. Spaatz's purpose was clear: he wanted to impose "heavy operational losses to their fighters." Spaatz's biographer David R. Mets noted that "USSTAF [United States Strategic Air Forces] was entering its most critical hour with a stable of aircraft that remain among the most esteemed in USAF history: the Fortress, Thunderbolt, Lightning, and Mustang." The USAAF was going to destroy the German aircraft industry, both in the sky and on the ground, and gain and maintain air superiority for the rest of the war.[47]

Fighter aircraft moved farther away from the bombers and engaged the Luftwaffe on their own terms, attacking from positions of advantage rather than starting most fights defensively. From one perspective, the bombers simply became the bait and enticed the Luftwaffe into coming up. The change in tactics worked. For the first time on 12 March 1944, fighter aircraft were out in front of the bombers and engaged the Luftwaffe's Messerschmitt fighters. The German air arm now faced an unwinnable dilemma: its fighters could no longer engage American fighters and bombers simultaneously, nor could they effectively fight the American fighters on their own. Throughout Big Week and into the spring of 1944, the Luftwaffe suffered dramatically. Although damage to German factories was limited and Germany's ability to produce aircraft continued to increase, the greater damage occurred with the loss of oil and gas and of trained pilots. The German fighting force lost nearly one-fifth of its pilots in February. While the Luftwaffe might be able to replace pilots, it could not replace their experience level.[48]

Germany was still engaged in a fight on the eastern front that kept a portion of the Luftwaffe tied to the action there. Most of the remaining fighters were committed to France and defense of the German homeland. These fighters were positioned to inflict as much damage as possible on the American bombers. Due in large part to the change in tactics that allowed American fighters to move away from the bomber formations, 56 percent of the German fighters in Germany and France were destroyed in the spring of 1944. The death of the Luftwaffe occurred in early 1944, and it was Army Air Forces fighters that killed it. The Luftwaffe became a broken organization. American airmen shot down hundreds of Luftwaffe fighters, but more importantly, Germany lost pilots it could not replace. Between the two-front war against the Soviet Union on one side and the Anglo-Americans on the other, the Luftwaffe could not replace and train new pilots fast enough. The balance of the air war tipped irreversibly in favor of the Allies.

It was not the strategic bombing of Germany's fighter industry that gave the Allies their great advantage in the air come 1944. Germany continued to produce more than enough aircraft to meet its needs, but there were fewer and fewer experienced pilots to fly these aircraft. Looking back to the early days of the strategic attacks in Europe, between January and June 1942, the Allies shot down just over 300 German fighter aircraft. In the same time frame of 1944, Germany lost nearly 2,900 fighters. Destruction of fighter aircraft, normally considered a tactical or perhaps operational objective, was the greatest contribution made by the 8th Fighter Command. The Luftwaffe had no choice but to try and interdict the bombers. The changes in doctrine and tactics gave the Allies a major advantage The USAAF seized the initiative, gained control of the skies over Europe, and never yielded.

As USAAF fighters systematically destroyed the Luftwaffe, bombardment emphasis shifted from deep strategic targets to the operational goal of preparing the battlefields of France for the Allied invasion. Destruction of the fighters was one piece of the air war in Europe in 1944; the other was the destruction of the transportation infrastructure in Germany and France. In advance of landings along the coast of France sometime between March and August 1944, the 8th and 15th Air Forces redirected their efforts toward support of the upcoming invasion. Targets included bridges, marshaling yards, rail lines, and, basically, any artery the Germans might use to send reinforcements toward the potential landing zones. The spring offensive brought movement in France to a virtual standstill.[49]

Taken together, the interdiction of the German Army's ability to reinforce itself once the invasion began and the systematic destruction of the Luftwaffe—which led to the gaining of air superiority—cannot be overstated in the Allies' successful invasion on 6 June 1944. The Luftwaffe was a shell of its former self with as few as 1,195 fighters "in a state of readiness" *in all theaters* in late May 1944. Despite this, the Allies turned their attention to attacking airfields in May 1944 to help ensure that the Germans could not bring the entirety of what remained of the Luftwaffe against the invasion force. The "airfield program" was conducted by the 9th Air Force using light and medium bombers, attack aircraft, and fighters. The 8th Air Force joined these attacks in the days leading up to the landings in France. The official history states it best: "As it turned out, German air opposition to the Normandy landings was astonishingly slight, far below the scale anticipated by the Allied air commanders. Indeed, one of the most remarkable facts of the entire war is that the Luftwaffe did not make a single daylight attack on D-Day against the Allied force in the channel or on the beaches." It became obvious in short order that the Allies were going to hold their positions and open the beachhead into the continent. However, the Normandy campaign dragged on for nearly two more months. The USAAF now turned its attention to aiding in breaking the Allied force out of the Normandy region.[50]

As the British forces pressed their attack on Caen, Omar Bradley launched Operation Cobra. On the morning of 25 July and lasting for an hour, 1,800 heavy bombers of the 8th Air Force relentlessly bombed a section of the St. Lo road 7,000 by 2,000 yards (roughly four miles long and three-quarter miles deep) in order to crack the German lines and force a breakout. This attack was preceded and followed by attacks by light and medium bombers and fighter and attack aircraft. This was truly an example of "carpet bombing," and in this particular case, it worked. The heavy bombers blasted a hole in the German lines and the US First Army began moving through it. Although the Germans continued to put up a tenacious resistance, their lines were no longer coherent; American units exploited the gaps and seams moving in and around them, flanking the German defenders. The Normandy campaign had its breakout point.

Historian Donald L. Miller called Cobra one of the "supreme military achievements of the European campaign." Still, the operation was far from a perfect example of using strategic bombers for tactical purposes. Spaatz was opposed to the operation, and with good reason. Although the attack achieved its goal, numerous bombs fell short, and for the second time in as many days killed over a hundred American soldiers, including Lt. Gen. Leslie McNair. Spaatz conceded after the attack that "short bombing, like short rounds from supporting artillery fire, was an unavoidable fact of life." Spaatz also noted that, unlike artillery, heavy bombers were not trained to work so close to the front lines and recommended that in the future they should not be used as replacements for traditional artillery attack. Eisenhower agreed saying, "I don't believe they can be used in support of ground forces. That's a job for artillery. I gave them a green light this time. But I promise you it's the last."[51]

Fratricide, or inadvertent bombardment of friendly troops, was not a new occurrence, and Spaatz must be given some leeway in his assessment, though harsh, that these occurrences were unavoidable. It was not a problem that would get better either with experience or technology. Friendly-fire incidents continued in America's more recent conflicts, and heavy bombers were used in close air support roles in Vietnam and Afghanistan as well. The contentiousness between the US Army and later the US Air Force over how to get close air support "right" has remained (some might say it's a festering sore) into the twenty-first century. Perhaps the fairest judgment on Cobra came from historian Richard Davis, who called the 8th Air Force's bombing mission that July day "its most effective and its most controversial."[52]

Return to Strategic Targets

Even through the Transportation Plan and the use of strategic bombers against tactical targets, Spaatz and his airmen remained focused on their core mission and their desire to cripple Germany's war-making ability. Despite the "Rise of Armageddon," as Michael Sherry termed it, American bombardment of aircraft factories and other targets, those on the ground did not experience Ragnarok. For the most part, it seemed as if German factories were little hindered by the American bombs raining down around them. German fighter production rose and continued to rise even as the CBO hit its full stride. From the third quarter of 1943 production rose by 30 percent. This trend continued in 1944. In the first quarter of 1944, German aircraft production doubled, by the second quarter, it tripled, and by the third quarter of 1944 the German aircraft companies were churning out 3,031 aircraft per month for a total of over

40,000 aircraft produced that year. In all, German industry produced three times as many weapons in 1944 as in 1941. Even with the numbers continuing to increase, production still fell well below what Germany needed to continue the war effort. Of course, aircraft production was not the problem; pilot "production" was. Ever since Doolittle turned his fighters lose against the Luftwaffe, the war in the air had taken a dramatic turn for the worse for the Axis. Truly, the Luftwaffe did not stand a chance by this point. Pulled in three different directions since 1942, the die was cast. In the East, the Red Air Force had more than regained its footing. In the south, North Africa, Sicily, and most of Italy had fallen. Now, in the west, the USAAF was pushing relentlessly toward the final destruction of the Luftwaffe and the collapse of the German state.

Ironically, despite gaining air superiority and demonstrating proof of the importance of fighter escort and the training and skills needed to engage with and destroy the enemy, the postwar Air Force chose to unlearn these lessons, and they would have to be relearned time and again in Korea and Vietnam. Rule one was gain and maintain air superiority. The efficacy of strategic bombardment theory in Europe will be addressed further at the end of this chapter, but suffice it to say, nearly eighty years after the end of the Second World War, the USAF and academic communities continue to struggle to determine what exactly the USAAF accomplished in the European theater in the war. The wholesale destruction of the Luftwaffe was clearly a positive byproduct of the aerial war against Germany and a success that aided in the Allies march toward Berlin. The *USSBS* conducted after the war stated, "Allied airpower was decisive in the war in Western Europe. Hindsight inevitably suggests that it might have been employed differently or better in some respects. Nevertheless, it was decisive." Ironically, it was not the industrial targets that the *USSBS* pointed to as a measure of this decisiveness; it was the air war against Germany's U-boats and the support airpower provided to the various Allied invasions that were the true measure of merit, both being items Allied air proponents argued against during the war.[53]

The Pacific Theater of Operations

The vastness of the Pacific theater in the Second World War necessitated it being broken down into several smaller theaters. Gen. Douglas MacArthur commanded the Southwest Pacific, and Adm. Chester Nimitz the Northern and Southern Pacific Ocean Areas. As such, airpower was divided not only between these two theaters, but also the China, Burma, and India (CBI) theater. Thus, from three directions, American airpower would slowly claw its way toward the Japanese mainland in arduous multiyear campaigns that brought Japan under attack from strategic bombardment. After the Doolittle Raid in

1942, it was not until 15 June 1944 that American bombs began falling on Japanese cities as part of a concerted effort that had been ongoing in Europe since 1942. Japan proved to be a more difficult enemy to approach.

The Southwest Pacific

Some of the more original uses of airpower occurred in the Southwest Pacific theater under the direction of George C. Kenney. Known as "MacArthur's Airman," Kenney was one of the truly innovative airmen of the war. Innovation is a term thrown around so much it has almost lost its meaning, but Kenney's employment of "skip bombing," his ability to deviate from prescribed doctrine, and his emphasis on "attack aviation" did much to harass the Japanese effort throughout the summer of 1942.

Kenney's efforts were indeed innovative. He made extensive use of all the aircraft available to him, including A-20s, B-25s, and B-26s. Kenney's forces attacked from all altitudes, but they made especially good use of low-altitude raids, a tactic that fit nicely to combat in the Pacific. This tactic involved using B-17s from low altitude to attack Japanese shipping. In one skip bombing incident, American A-20s mistakenly attacked several Australian barges. MacArthur told Kenney to find out what had happened. When Kenney located the responsible crews, he told them, "Australians should not bomb American barges any more than American pilots should bomb Australian boats."[54] He went on to tell the crews, "I explained to them that what pained me more than anything else was the fact that they missed the boat with all their bombs. I therefore considered that they were not fit to belong to any outfit in the Allied Air Force and sentenced them to daily, morning and afternoon, bombing practice ... until they could hit something with a bomb besides the Solomon Sea."[55]

One of Kenney's subordinates was Brig. Gen. Kenneth Walker, who had served on the AWPD-1 development staff along with George, Kuter, and Hansell. In the Pacific, Walker commanded the V Bomber Command. Kenney called him the "best bombardment commander I had" and ordered him to stop flying missions for the dual purpose of keeping him alive and allowing his crews to conduct their missions without having a brigadier general getting in the way. Kenney was rather emphatic about this second point and noted that if something went wrong on a mission, Walker would just be "in the way." Walker disobeyed Kenney's orders and was killed in battle on 5 January 1943.[56]

A crewmember on another B-17 left this remembrance of the mission where Walker did not return:

> "Hey, there's somebody in trouble behind us." So we made a turn and looked back and here was an airplane, one of our airplanes, going down, smoking and

on fire, not necessarily fire, but smoke anyway, and headed down and obviously headed for a cloud bank with the whole cloud of fighters on top of him. There must have been about fifteen or twenty fighters. Of course, they gang up on a cripple, you know, polish that one off with no trouble, but he disappeared into a cloud bank and we never saw him again. It turns out it was the general.[57]

Kenney planned to reprimand Walker if he was found alive. When he informed MacArthur of this, the general laconically said, "All right George, but if he doesn't come back, I'm going to send his name in to Washington recommending him for the Congressional Medal of Honor." President Roosevelt later awarded Walker the medal posthumously. Walker represented the maverick streak so prevalent in American airmen dating back to Billy Mitchell. Despite orders to the contrary and these missions taking a physical and emotional toll on him, Walker continued to fly. His spirit and determination to continue leading his men at all costs mark him as a truly great airman of an aerial way of war.[58]

In March 1942, at the Battle of the Bismarck Sea, Kenney's forces dealt a significant blow to a Japanese convoy on the way to Lae, New Guinea. The Japanese convoy lost between eight to fourteen transports and merchant vessels, four destroyers, and nearly three thousand men. Kenney noted in his diary: "At three o'clock in the morning I woke up General MacArthur to give him the final score. . . . I had never seen him so jubilant."[59]

Flying the Hump

Another maverick left his mark in the war against Japan, but thousands of miles away from Walker and the Southwest Pacific. Those who read popular histories tend to be most familiar with the European and Pacific theaters of the war. Too often overlooked in the histories of World War II is the China, Burma, and India theater. Connecting the CBI through a perilous air route was "the Hump." Flying from India into China and over the Himalayas was known as "flying the Hump." The route served the purpose of resupplying the Chinese army in its fight against Japan. Later it provided supplies necessary for the conduct of operations by the 20th Air Force.

The importance of "flying the Hump" is difficult to overstate. From January 1943 through August 1945, 685,000 tons of cargo moved along this particular aerial route. Of that total, nearly 400,000 tons was gasoline and oil. This airlift was accomplished with a fleet of aircraft that rarely exceeded two hundred in 1943–44. Beginning in November 1944, with the war in Europe inching toward its inexorable climax, things began to change in the CBI. More aircraft arrived by the month, and the gross tonnage being flown over the mountains increased drastically. This was due not only to the availability of resources, but also to the singular drive of Brig. Gen. William H. Tunner. The vast majority of the supply runs occurred under Tunner's direction.[60]

Tunner did not arrive in the CBI theater until September 1944. His predecessor was Brig. Gen. Thomas O. Hardin. Although Hardin had worked hard to increase the tonnage of supplies getting over the Hump, an associated increase in accidents and lost aircrews occurred on his watch. Tunner was sent in to keep the tonnage rising but, at the same time, to cut down on the accident rate. Although Tunner's autobiography is somewhat self-serving, he was entirely correct in calling this particular assignment the "graveyard for commanders." It was not an assignment he relished taking on. In Tunner's perspective, the CBI was filled with exiled officers, alcoholics, and misfits. Still, he took over the command and increased not only safety but morale as well for the officers who felt they had been consigned to the backwater of the war. Tunner was later criticized by Lieutenant General Stratemeyer for becoming too safety-conscious and letting his accident rate fall behind that of the combat squadrons. Tunner found this ridiculous. In a rare self-effacing comment, Tunner noted that he might have increased safety and morale but laid the increase in tonnage at his predecessor's feet, saying that to Hardin "must go much of the credit for this tonnage."[61]

The operations in the CBI and flying the Hump became a mainstay in the study of airlift operations. Perhaps less well known than the forthcoming Berlin Airlift, it nevertheless demonstrated the significant importance of cargo and transport operations. Without the men flying the Hump and providing the logistic tail, initial bombing operations as part of Operation Matterhorn, the name given to the bombing of Japan from China, would not have been possible. Thus, the airlift function should be viewed as equally emblematic of an air force way of war as are fighter and bomber actions. Airlift operations do not necessarily have the popular appeal of dogfights and bombing runs, but they were equally as dangerous and contributed as much, if not more, to the success of the Army Air Forces during the war. Historian Daniel Haulman noted of the Hump missions: "The Army Air Forces in the China-Burma-India Theater succeeded in keeping China in the war, defending India, and liberating Burma. The accomplishments of American airmen in the China-Burma-India Theater were both legion and legendary."[62]

The Strategic Air War against Japan

The strategic bombing campaign against Japan altered traditional concepts of American bombing doctrine more than the leaders of the USAAF expected. Throughout the war in Europe, it was understood and expected that American bomber formations would attack from *high altitudes*, using *precision bombing* methods, and do so during daylight hours. HADPB was the way of business in Europe. Conceived in the wake of the First World War, indoctrinated at ACTS, and used against industrial targets in the cities of Europe, it was understood to be the way of war for the USAAF. With the introduction of the B-29, several factors forced a change in this method of employment. Before that could be accomplished, the Marine Corps, Navy, and Army had to get airfields close enough, with long enough runways, to Japan to launch these raids.

These first attacks by bombers from the 20th Air Force under the command of Haywood Hansell flew not from any Pacific islands, but from the Chengtu (Chengdu) Valley in central China in June 1944. These were the first raids against the Japanese homeland since the Doolittle Raid in 1942. For perspective, the raids against Japan began the same month as the Normandy invasion. By this point, Germany had been under sustained bombardment from Allied forces for two years. Attacks from the XXI Bomber Command in the Marianas began in October 1944.[63]

It was time to put the technological wonder that was the B-29 to work. Initially conceived of in 1939 and making its maiden flight in 1942, the B-29 Superfortress (described in the intro to Epoch II) was significantly different from its predecessor heavy bomber brethren the B-17 and B-24. It was twenty feet longer than the B-17 tip to tail and had a wingspan more than forty feet longer than the B-17's. In every other aspect, it was a leap over the B-17—it could go faster, climb higher, and was in any other area of comparison a better strategic bomber. The B-29, depending on mission length and altitude, could carry anywhere between 500 and 14,000 pounds more bombs than the B-17. Its maximum load was 20,000 pounds. Its real technological advancement was in its pressurized cabin, which allowed crew members to move about and not be tied into an oxygen system. Its range was a manifold jump over the B-17's.[64]

Throughout its development, production, and deployment to the CBI and the Pacific, General Arnold maintained a tight organizational control over the aircraft. The 20th Air Force "operated under the direct command of General Arnold, as executive agent of the Joint Chiefs of Staff, and Commanding General of the AAF." Arnold's sole purpose in maintaining control was to ensure that the B-29 conducted only strategic bombardment missions and to not let it be siphoned off for other missions, as happened too often with bombers in the ETO.[65]

Once B-29s began bombing the Japanese homeland, however, several problems developed. For starters, the B-29s flew at higher altitudes, which complicated the precision problem. The higher a bomber flew, the less accurate it was. Second, there were troubles with navigation, fuel consumption, and the weather.

Third, the high altitude worked against sensitive equipment. Although the B-29 was pressurized, equipment still tended to freeze and malfunction. All of these factors worked against the B-29 crews. The B-29 might be a technological wonder, but it was having operational growing pains.[66]

There was also the matter of attacking targets on the Japanese home islands. Although the kamikaze raids against Navy ships are well known, the Japanese employed similar tactics against the B-29 bomber streams. In one incident a Japanese fighter, smoking badly and slowly drifting downward struck a B-29 but failed to do enough damage to bring down the bomber. Moments later, a second suicidal attack by a fighter brought down the Superfortress.[67]

Although several proponents of American strategic bombing theory opposed the use of B-29s as aerial mining platforms against Japanese harbors as part of Operation Starvation, it proved to be more successful than the ongoing high-altitude bombing missions. Japan was now subject not only to high-altitude bombing by the B-29s, but also to the complete strangulation of whatever supply lines remained open to the island nation. The noose had finally tightened around the home islands.

Haywood Hansell bore the brunt of the unknowns that challenged the 20th Air Force: the jet stream, untrained aircrews, a technologically advanced and sophisticated aircraft entering combat for the first time, bombing accuracy, and suicidal attacks from the remnants of Japan's air force. Hansell set about correcting these problems, but he needed time to work these issues and achieve the results Arnold wanted. Arnold was not willing to wait, and in January 1945, he fired Hansell; however, rather than do so personally, he sent Maj. Gen. Lauris Norstad to handle it—a fact both Norstad and Major General Kuter found distasteful.[68]

The man who replaced Hansell in January 1945 was none other than Curtis LeMay. After his work with the 8th Air Force and the air war in Europe, LeMay moved over to command the XX Bomber Command in August 1944 and the bombing of Japan from the CBI. Arnold fired Hansell and raised LeMay, who now directed operations for both XX and XXI Bomber Commands. LeMay now rapidly got over his former adherence to high-altitude daylight bombing. Simply put, LeMay abandoned USAAF doctrine, choosing instead to expand on a tactic Hansell had used on two occasions: low-altitude missions using incendiary devices.[69]

LeMay rapidly overcame the long-held principles of high-altitude strategic daylight bombing. LeMay proposed to go in low and at night. This represented a change of both tactics and strategy. In March 1945, the B-29s firebombed Tokyo for the first time as part of Operation Meetinghouse. Beginning that month, the doctrinal changes went into effect and were made permanent. As far as LeMay was concerned, HADPB was dead. The lower altitudes (below 7,000 feet) of the incendiary raids increased the bomb payload in each aircraft. The Japanese air defenses and lack of night fighters allowed the bombers to attack at will without fear of significant opposition. LeMay's 20th Air Force targeted four cities: Tokyo, Nagoya, Osaka, and Kobe.

According to the *USSBS* written after the war, in the first raid, "fifteen square miles of Tokyo's most densely populated area were burned to the ground. The weight and intensity of this attack caught the Japanese by surprise." The low-altitude approach also fixed the accuracy problem of the B-29. Later reports noted that between 35 and 40 percent of bombs landed within 1,000 feet of their intended targets. With the use of incendiaries, the targets and all surrounding areas were destroyed. The now well-known results proved catastrophic to those on the ground. One hundred thousand dead and fifteen square miles of Tokyo burned to the ground. For miles in every direction, ash settled to the ground, inches deep in some places, leading the Japanese to call the attack the "night of the black snow."[70]

In the years following the Second World War, but more pointedly since a reexamination of the war began in the 1990s, some have questioned whether the incendiary raids against Japan and the use of the two atomic weapons count as war crimes. Major General Kuter, who was only in the Pacific Ocean Area briefly, reflected on the fire-bombing raids years later saying, "Japan burned so violently, and I'll agree that we may have had more humane considerations with our European ancestors on the European side than we did with the Orientals. Burning a city was looked at as a military attainment, and we weren't thinking about the thousands of women and children in the city. I did after I was stationed there—later on."[71]

Regarding the effects of the bombing of Japanese civilians, the *USSBS* concluded in these stark and dramatic terms: "Total civilian casualties in Japan, as a result of 9 months of air attack, including those from the atomic bombs, were approximately 806,000. Of these, approximately 330,000 were fatalities. These casualties probably exceeded Japan's combat casualties which the Japanese estimate as having totaled approximately 780,000 during the entire war. The principal cause of civilian death or injury was burns. Of the total casualties approximately 185,000 were suffered in the initial attack on Tokyo of 9 March 1945."[72]

From June 1944 to August 1945, the US Army Air Forces methodically laid waste to Japanese cities in addition to dropping mines in Japanese harbors as part of Operation Starvation. Targets included Tokyo, Kobe, Kawasaki, Osaka, and dozens of other cities. Total estimates of Japanese killed range from the 800,000 noted above to more than a million. These bombing raids ended in early August 1945 with the atom-bombing of Hiroshima followed by Nagasaki three days later.

Much ink has been spilled over the "special missions" of the 509th Composite Bomb Group and the entrance of atomic warfare into the annals of history. Was their use necessary? Did they hasten the end of the war? Would the Japanese have surrendered without their use? Counterfactuals are impossible, but the *USSBS* and the remembrances of senior leaders and other historians offer some interesting conclusions that differ significantly from their findings in the ETO. The surveys noted that "Japan would have surrendered even if the atomic bombs had not been dropped," but they also echoed Arnold's thoughts that they would do so "even if no invasion had been planned or contemplated."[73]

Spaatz stated years later, "I think the Japanese would have surrendered within a month or two from the time they did surrender if we had used conventional bombing. General Arnold used this argument on the Joint Chiefs of Staff, that is, to bomb without the use of the Atomic Bomb." This was strictly true, but Arnold's argument was not out of any moralistic necessity to saving the lives of Japanese citizens; he wanted to prove that airpower could force Japan to capitulate.[74]

Historian David Mets noted, "There was some sentiment among airmen that the nuclear weapons ought not be used. Arnold is recorded as having raised doubts about the wisdom of doing so, at least partly because he felt that the conventional bombing campaign combined with the blockade would be sufficient to bring about Japan's capitulation without the costly invasion of the Japanese homeland and the attendant causalities."[75] In other words, Arnold wanted to continue to use his bombers without atomic weapons to prove their efficacy.

Spaatz also thought there was another answer short of destroying another Japanese city: "I thought that if we were going to drop the atomic bomb, drop it on the outskirts—say in Tokyo Bay—so that the effects would not be as devastating to the city and the people. I made this suggestion over the phone between the Hiroshima and Nagasaki bombings, and I was told to go ahead with our targets."[76]

Conclusion

The authors of the USSBS also took the time to put the effects of bombing civilians into the larger context of the overall bombing campaigns, and they came to the following conclusion: "The mental reaction of the German people to air attack is significant. Under ruthless Nazi control they showed surprising resistance to the terror and hardships of repeated air attack, to the destruction of their homes and belongings, and to the conditions under which they were reduced to live. Their morale, their belief in ultimate victory or satisfactory compromise, and their confidence in their leaders declined, but they continued to work effectively as long as the physical means of production remained. The power of a police state over its people cannot be underestimated."[77]

In this, the surveys' authors could not have been more wrong. The "power of a police state" has little to do with the human response to tragic events. Did the Soviet people on the eastern front endure hardships because of a police state? People throughout history have displayed remarkable coping mechanisms including the ability to continue their day-to-day lives in the face of aerial bombardment or siege warfare. People's desire to continue living their lives can be traced more to a desire for survival and a return to normalcy than it can be linked to belief in and support of an authoritarian regime.

Two questions remain with regard to the air force's identity and style of war at the end of the Second World War: What did World War II do to the American way of war for airmen themselves? What did World War II mean for the Air Force as an institution going forward?

The first question seems easier to answer. The Second World War began with an Army Air Corps decimated in the PTO and virtually nonexistent in the ETO. On the North American continent, the Army Air Corps, thanks in no small part to the full-time manufacturing capacity of the United States, developed into the Army Air Forces with the most sophisticated and specialized aircraft known to man. While debates continue to rage about the best fighter or bomber of the Second World War, what is not in debate is the sheer size and technological advancements of aircraft made during the war. Across the spectrum of aircraft types, from fighters, bombers (heavy, medium, and light), attack aircraft, to cargo and transports; from single-seat/single-engine to multiengine, pressurized bombers, the Second World War witnessed a dramatic evolution in aircraft development.

The Army Air Forces enjoyed other advantages during the war. These included the isolation of the continent from attack and a training pipeline that produced pilots better prepared for combat than their counterparts in Germany and Japan. The significance of pilot training cannot be overlooked. By the end of the war, America had a surplus of pilots. The fact that they were better trained than their adversary is also important: training for combat—and this included basic flight school through advanced tactics and employment—is as emblematic of a way of war as is the ace pilot or steely-eyed leader of a bomber formation.

In total, 141,387 American aircrew members were either killed, wounded, or missing in action or captured as prisoners of war. It was a tremendously high casualty rate. Even when compared to the vast numbers of American casualties across-the-board during the war, it was an especially high rate. Historian Mark

Wells said the bomber crews continued to fly day in, day out because "the principal reason most airmen stuck things out in combat was because of the spirit of cohesion and teamwork that permeated units and individual aircrews."[78]

The United States Army Air Forces did all that was asked of it and more. This is not hyperbole. The USAAF's achievements owe as much to the "arsenal of democracy" as to the leaders and airmen who flew the missions. Strategic bombardment against the enemy's homeland, fighter and escort missions, close air support for ground troops, cargo transport—the Army Air Forces did it all. To accomplish offensives simultaneously in two global theaters, it relied on the means and resources provided by a dynamic national economy, and on a training pipeline that was second to no other nation's—as well as on men and women who performed the needed roles and missions.

As to what World War II meant for the Air Force as an institution going forward? The independent Air Force that emerged shortly after the end of the war hung its hat on the continued use of strategic bombardment in both doctrine and theory. This, of course, was predicated on the fact that in the immediate aftermath of the Second World War, the United States Army Air Forces was the *only* organization capable of carrying and delivering atomic weapons. With no other service or branch capable of atomic delivery, the Air Forces had finally found a true reason for independence. The training and equipping of a nuclear air force could not possibly be left nestled within the United States Army. At last, the Air Forces had a mission that clearly necessitated independence and separation from the Army. For much of the next thirty years, strategic bombardment and, more specifically, the delivery of atomic weapons became not only the mission of the Air Force, but it truly dominated force structure and funding lines for the American military as a whole. When the USAAF became the United States Air Force in 1947, it remained the only military force capable of delivering atomic weapons.

The Cold War began shortly after the end of the Second World War. The United States held a monopoly in atomic weapons for at least a few years. The USAF's capability to deliver atomic weapons was centered exclusively in the Strategic Air Command. For this reason, the Strategic Air Command virtually ruled the Air Force and the Department of Defense.

CHAPTER 4

A Strategic Air Force

The jetomic age burst upon the world. Never before has man been subjected to such rapid and revolutionary technological change.

—Lt. Gen. Laurence S. Kuter, 1954

The US Army Air Forces ended the Second World War postured well for future conflict. It was, at the time, the largest and most modern air arm in the history of the world. The advent and use of atomic weapons clearly demonstrated the absolute power that air warfare could bring to bear. Although the effectiveness of the US strategic bombing campaigns continues to be debated by historians, there was no doubt in 1945 that the Army Air Forces had contributed significantly to the defeat of Nazi Germany and, more so, of Imperial Japan. The postwar Air Force gained its independence and despite a massive drawdown remained trained, committed, and prepared to conduct operations, which represented a new construct for an American fighting force. The leaders of this newly minted United States Air Force had each proved themselves in the crucible of global war.

The postwar Air Force, built around the strategic imperative of delivery of atomic weapons, also oversaw a technological shift that was every bit as revolutionary as during the interwar period from 1918 to 1941. This time, advances in aeronautics, primarily in the fields of aerodynamics and jet engines led to improved developments in the application of airpower in the form of aerial resupply and logistics, air-to-air refueling, and in the designs of the combat aircraft themselves. No other time in the history of airpower development saw as much change as the post–World War II period. Between 1945 and 1965, the Air Force procured jets and bombers at an amazing rate. Advances occurred at such a rapid pace that some aircraft were becoming obsolete by the follow-on generation of aircraft developments even as they rolled off the assembly line. Some of these aircraft, created for a singular purpose, found that operational requirements altered their intended mission. In the coming decades, bombers intended for nuclear delivery would provide adequate close air support. Fighters intended to deliver nuclear weapons would become conventional attack aircraft. More than innovation, operational necessity required rapid adaptation. That was still in the future, though. Of immediate concern was independence and the stand-up of the nation's newest armed service.

In February 1946, Gen. Carl Spaatz became commanding general of the United States Army Air Forces, succeeding Gen. Hap Arnold following the latter's retirement that same month. There was no shortage of advice given to Spaatz, including from his old boss Arnold, who wrote a lengthy letter to him shortly after the end of the Second World War. The letter listed twelve areas of concern that Arnold wanted Spaatz to consider: "discipline, your staff, advisory council, weeding out misfits, unfits and the partially fits, scientific development, Dr. Von Karman's report, Air Transport Command, appropriations and allocations of funds, flying pay, mail list, housing, and extracurricular activities." A seemingly scatter-shot list, Arnold's letter to Spaatz is instructive as to the transition of power from the man who led the Army Air Forces throughout the war to the man who would lead it as an independent United States Air Force. While some topics do not merit much attention, others demonstrate in just what direction Arnold thought the "Air Forces," as he called them, should go in the future. Some of the more mundane topics include Arnold's desire to get rid of "flying pay" for pilots, which he felt created a "thorn in the side of ground soldiers, of legislators, and many others." To that end, Arnold thought Spaatz should "delete completely increased pay and above everything else, the term 'Flying Pay.'" Arnold also cautioned Spaatz not to allow cliques to develop among senior officers, warning him that "even at this writing there inclined to be cliques—the old college tie clique, the European clique, the Mediterranean clique, the Pacific clique—and no organization can operate with cliques." The Air Force has in fact spent much of its existence fighting against the cliques and "stovepipes" that have proved to be an enduring problem.[1]

However, in this letter Arnold directed more attention and time to ensuring that Spaatz recognized the need for continued technological development in the Army Air Forces, already a major focus for the organization and an idea that in the future became synonymous with the Air Force itself. Under scientific development, Arnold stressed the importance of maintaining a technological edge in the future: "We [must] find ways and means of earmarking certain bright and outstanding young men in technical colleges such as

M. I. T. and Cal-Tech for commissioning in the Air Forces. We could tell them that here is a career job and that hereafter science and research will have the same relative importance as pilot training—piloting of aircraft—and that there is a future in the Air Forces for these men.... We must have good scientific and technical officers if we are to maintain our position as the most powerful Air Force in the world." Arnold also noted that the USAAF needed the very best scientific minds for the "future development of Air Forces gadgets and techniques."[2] Spaatz heeded much of Arnold's advice. Of course, he planned to put his own mark on the new service, but chief among Spaatz's initiatives was to "press scientific research and development and ensure planning for the industrial bases of airpower."[3]

Henry "Hap" Arnold, five stars adorning his shoulders, General of the Army, and later General of the Air Force (the only officer to hold that particular distinction), retired from active-duty service on 30 June 1946. Arnold and his wife, Bee, retired to Southern California. Already prone to health problems—he had four heart attacks during the war—Arnold set out to write his memoirs. He eventually published *Global Mission*, though he suffered another heart attack in the process. In retirement, he even lent his name to the Pabst Brewing Company for an advertisement calling Pabst Blue Ribbon "the finest served anywhere!" He died in 1950, one of the first to pass of the leaders who oversaw the US military during World War II.

Spaatz became the first chief of staff of the USAF on 18 September 1947. The United States Air Force came into existence as an independent armed service as part of the National Security Act of 1947; for the time being, it held the truly independent mission of delivery of atomic weapons. Thus, 18 September 1947 marks not only the USAF's birthday, but also its "Independence Day." Spaatz had under his command a group of seasoned combat veterans and a cadre of officers with global experience thanks to their service in the war. This list included such notables as George Kenney, Larry Kuter, Curtis LeMay, Pete Quesada, George Stratemeyer, and Hoyt Vandenberg. The American public knew these men almost as well as Babe Ruth had been known in the 1920s and '30s. A number of them had even appeared on the cover of *Time* magazine.

The Early Air Force

In the year prior to independence, the soon to be defunct US Army Air Forces divided itself into major commands (MAJCOMs). On the home front, there were four (each headed by a general officer): Air Defense Command (under George Stratemeyer), Air Training Command (John K. Cannon), Strategic Air Command (George Kenney), and Tactical Air Command (Pete Quesada). Since SAC was the only command capable of nuclear delivery, it made sense that SAC garnered a larger share of the resources. In 1946, personnel assignments demonstrated just how much larger SAC was than the other commands. Upon their creation, the three principal nontraining commands totaled just over 117,000 Army Air Forces personnel. SAC had 84,231 members; TAC, 26,000; and Air Defense Command (ADC), a paltry 7,000. This proved how much the USAF aligned its identity and force structure with SAC.[4]

SAC's dominance grew, in no small part, to the machinations and leadership of Curtis LeMay once he became its head. The newly created United States Air Force built dominance over the Department of Defense, and all aspects of preparing for combat took a back seat to "strategic dominance." The Air Force built newer and more technologically advanced nuclear delivery bombers but did not want to see them used in any proxy war, preferring to keep them ready for the—perceived then as probable—coming war with the Soviet Union. Indeed, LeMay and SAC considered that they were already at war. LeMay said of this mentality: "We had to operate every day as if we were at war, so if the whistle blew we would be doing the same things that we were doing with the same people and the same methods." Beyond an ever-growing strategic bomber fleet, the USAF built interceptor aircraft that could be either outfitted as nuclear delivery aircraft themselves or specifically designed to shoot down incoming Soviet bombers. Finally, by 1957 the Air Force had embraced the creation and fielding of an intercontinental ballistic missile force. As a result, the Air Force allowed many of the tactical advantages learned during combat in the Second World War to atrophy. First, however, the Air Force had to contend with air transport, and not bombardment forces, as the unexpected face of the USAF.[5]

MATS and the Berlin Airlift

It is somewhat ironic that after four years of employing strategic bombing as the primary method of airpower operations during the Second World War, and after the establishment of the independent US Air Force with its dominant Strategic Air Command, it was actually transport aircraft that became the visible representation of airpower during the early Cold War. This was not missed by Hap Arnold, who had presciently noted in his final letter to Spaatz that

> the technique, knowledge of procedure and experience that we have learned in the Air Transport Command should never be lost. Accordingly, it would seem to me that we should have in peace time an Air Transport Command running routine services between the United States and our bases in the Azores, Iceland, Greenland, Alaska, Okinawa and the Philippines. . . . The size of the Air Transport Command should be such that it, together with its reserve in the airlines

themselves, can pick up and carry one Army Corps and transport same to either Alaska or Iceland. It is believed a thousand planes would be required for such a transfer of personnel.[6]

The Army Air Forces, and later USAF, already had an existing Air Transport Command (ATC). The ATC operated throughout the Second World War comprising six domestic ferrying groups and a foreign command running between theaters. The ATC was responsible for all ferrying activities, establishment of intertheater air routes, and transport of all military personnel by air across the globe (excluding the movement of combat troops, such as the airborne divisions that fell under the troop carrier units).[7]

In June 1948, the ATC was merged with the Naval Air Transport Service (NATS) to create the Military Air Transport Service (MATS), commanded by Maj. Gen. Laurence Kuter. Although principally remembered for his contributions to bombardment theory and the development of AWPD-1 and FM 100-20, Kuter had moved over to Air Transport Command at the end of the war. The development of a larger and more effective airlift program that combined Navy and Air Force assets under the same umbrella paid dividends almost immediately. When Joseph Stalin ordered access to West Berlin blockaded on the ground less than a month after the establishment of MATS, the United States and the United Kingdom moved to use airpower and aerial transport as a means of keeping the city's Allied zones supplied.

Stalin did not initiate a blockade of the city's Western-occupied zones with the express purpose of making the Americans, British, and French remove their forces from the city and leave it in the control of the Soviets. Rather, it was a move by Stalin and the Soviet Union calculated to force the Western Allies to renegotiate on currency in Germany. If they left the city anyway, then so much the better, Stalin figured. Instead, Stalin's blockade inadvertently "became a propaganda fiasco and strategic failure."[8]

What history remembers as the "Berlin Airlift" was actually a mission that stretched across the globe and was not localized to the German city that bears its name. MATS aircraft and personnel traveled from as far away as Hawaii to participate. Kuter ordered his deputy, Maj. Gen. William Tunner, to report to Europe and take command of the airlift operations. Although Tunner was the MATS deputy, he headed for the headquarters of the United States Air Forces in Europe (USAFE), where the airlift was already underway under the name "Operation Vittles." USAFE's principal purpose was to block any perceived Soviet threat to Western Europe. Most Air Force leaders believed a response to any Soviet threat meant fighters and bombers and not necessarily large numbers of air transport aircraft. This stance immediately changed with the beginning of the blockade. The airlift itself lasted 464 days and demonstrated America's commitment to Western Europe. More than 2 million tons of supplies were flown into Berlin by the British and American forces.

It was a complete failure for Stalin, who did not believe it was possible to keep the city supplied using only aerial resupply. Historian Daniel Harrington, author of *Berlin on the Brink*, stated that Stalin "watched dumbfounded as the airlift circumvented his blockade." The Soviet leader was not alone in his doubts, for the airlift exceeded everyone's expectations. Historian Robert C. Owen concluded, "Operation Vittles humiliated the Soviets and forced them to back down from their goal of starving Berlin into submission. Within a few weeks of slamming shut the gates into the city, they became unhappy witnesses to a growing stream of transports that undermined and eventually made a mockery of their strategy."[9]

In addition, there was the public relations victory thanks to "The Candy Bomber," Gail S. Halvorsen and his C-54. History remembers Halvorsen for dropping bars of chocolate by parachute to the children of West Berlin from his C-54. Thanks to Halvorsen, the C-54 became emblematic of the airlift and thus of America's (and the USAF's) airpower. The USAF flew 189,960 missions and delivered more than 1.68 million tons of cargo to Berlin over a duration of fifteen months. Militarily, culturally, and ideologically speaking, the Berlin Airlift was a major success for the newly created United States Air Force. Despite the great public relations victory, attention after the airlift returned to the very real threat posed by a possible attack from the Soviet Union against the homeland of the United States. Unlike previous conflicts, America's Cold War enemy had the ability to reach across oceans and polar ice caps to attack the "home field" of the United States.[10]

Dawn of the Jet Age

It is common to speak of "generations" when discussing fighter aircraft. For example, current F-22 and F-35 represent the "fifth generation" of fighters. The concept of jet generations did not appear until the development of the F-22 in the 1990s, primarily because the Lockheed Martin Corporation used it as a messaging tool to differentiate the F-22 from every previous fighter. Thus, the developers and fliers of the USAF's first generation of jet aircraft did not recognize they were operating in the "first-gen" arena. They did, however, certainly recognize that the aircraft they were piloting—and this includes everything from the Me-262 and P-80s to F-86s—altered the very speed, if not the nature, of air-to-air combat.[11]

The development of jet aircraft for the Army Air Forces began during the Second World War. The first jet, the P-80 Shooting Star, had its maiden flight on 1 June 1944, just days before the Normandy invasion, but it did not enter production until the end of 1945; when the war ended that September, so did much of the P-80s initial production. It did not enter operational service until 1946. By this time, "pursuit" had given way to "fighter" for designation purposes, so the P-80 became the F-80. The F-80 saw service in Korea but was easily outclassed by the MiG-15. The same was true of the F-84 Thunderjet. Envisioned as an air-to-air fighter, the F-84 became an excellent interdiction aircraft functioning in the air-to-ground role. The F-84 was also the first aircraft selected for use in the Air Force's Aerial Demonstration Team, the Thunderbirds.[12]

The USAF solved the problem of the MiG-15 with the introduction of the iconic F-86 Sabre. Following these earliest fighters came such lesser-knowns as the F-89 Scorpion and the F-94 Starfire, both of which saw widespread use in the interceptor role for the ADC. This closed out the first and second generations of jet aircraft used by the USAF. Then came what is now considered the third generation, more commonly known as the Century Series.

The Bombers

Bombers developed along a similar path to the "generations" concept of fighter aircraft. Even more so than the fighters of the newly created Air Force, it was bombers that secured independence for the air arm—that is, B-29 bombers capable of carrying the early atomic bombs, followed in succession by the B-36 through the B-52. The bombers of Strategic Air Command became the raison d'être for the USAF.

The first of the post–World War II bombers to enter wide service was the B-36 Peacemaker. Design of the aircraft had begun well before the United States entered World War II, but it was not ready for service until 30 August 1947 and was not accepted by the Air Force until June 1948. The B-36 dwarfed its predecessor the B-29, which, as discussed in the previous chapter, was itself a revolutionary advancement over the B-17. However, the B-36 was simply a Frankenstein's monster of an aircraft. It was—in a word—gigantic. At 162 feet long with a 230-foot wingspan, it remains the largest combat aircraft ever built. Six rear-facing turboprop engines provided only a portion of the power that an additional four jet engines set on the outboard portion of the wings contributed. This led Convair aerospace officials to state the aircraft had "six turning, four burning," although mechanics and pilots derisively said the plane had "two turning, two burning, two smoking, two joking, and two missing." The Peacemaker served one purpose: the delivery of any atomic weapon in the USAF's arsenal at the time. Although it can create cognitive dissonance to one's modern sensibilities, the use of atomic or nuclear weapons made perfect sense to Air Force planners and senior leaders in the early days of the Cold War.

Coming along at the dawn of the jet age, the B-36 was arguably obsolete by the time it entered active service. The Peacemaker was thereafter followed by the introduction of all-jet bombers. Turboprops gave way to turbojets, which increased every aspect of the bombers capabilities: range, payload, and so on. The mainstay of SAC's nuclear delivery in the 1950s and 1960s was the B-47.[13]

Both the B-36 and the B-47 featured prominently in the film *Strategic Air Command*, which starred Jimmy Stewart. Made with the full approval and cooperation of the Air Force, the movie was a hit. It truthfully, but sympathetically (some might argue, propagandistically), portrayed life inside SAC in the early 1950s. It brought the life of aircraft and crewmembers, but more importantly, of the Strategic Air Command itself, before viewers, highlighting the important and no-fail mission of the command. The commander of SAC in the film, General Hawkes, is a stand-in for SAC's real commander at the time, World War II hero Curtis LeMay, right down to the clenched cigar. The film medium proved an effective way to translate the Air Force's mission, its way of war, into a palatable and digestible format that the American people could not only understand, but also identify with.

The B-52 Stratofortress replaced both the B-36 and the B-47; the B-52 eventually became the longest-serving bomber in the Air Force's history. In total, the USAF received a total of 742 aircraft in six different models (A–F). The Air Force officially credits the B-52 with being "the backbone of the manned strategic bomber force for the United States." More than sixty years after entering operational service, and with no retirement in sight, seventy-six remain in service either with the active force or in the reserves. Although conceived of as a strategic bomber for the delivery of SAC's nuclear arsenal, the B-52 proved to be adept at other missions as well, including conventional bombing. In Vietnam, it also provided close air support to the Marines at Khe Sanh and participated in Operations Desert Storm, Allied Force, Enduring Freedom, and Iraqi Freedom. It is the only aircraft to have seen combat in all of those operations.[14]

Air Defense and the Century Series

As the Cold War continued, the US Air Force shaped itself not only around SAC, but also around the mission to defend the continental United States against Soviet attack. Alongside SAC as the dominant major command inside the USAF, there developed a different type of air force and a different conception of how to use that air force that has been historically overlooked: that of an air defense force to protect the homeland from incoming Soviet attack.

When the Cold War began, or gradually escalated as the case may be, the American military, and the USAF in particular, started planning for and developing a "defensive air shield" to be used to locate, track, target, and destroy incoming Soviet bombers.[15] When the Air Force celebrated its independence as a separate service in September 1947, it was understood that the new branch would take the lead both in strategic attack missions and in defending the homeland—and just as importantly, by providing adequate warning to SAC—from aerial bombardment.[16]

Thus, as postwar reorganization took shape, another new major command, Air Defense Command, began operations. The command was established in 1946, and it became a wholly separate and equal major command in 1951, headquartered at Ent Air Force Base, Colorado. The ADC then divided its forces regionally, assigning each a section of the United States to protect.[17]

In 1954, the other armed services added their defensive assets into the fold and stood-up a new a multiservice unified command: the Continental Air Defense Command (CONAD). But the ADC continued to act as the Air Force arm of this new joint command, or what came to be called Geographic Combatant Commands (GCCs). CONAD defended the homeland. Included in the CONAD mix were Army Antiaircraft Command (ARAACOM) and Naval Forces CONAD (NAVFORCONAD). The late 1950s also saw the United States and Canada working closely together in the realm of air defense of North America, resulting in creation of the North American Air Defense Command (NORAD) in 1957. The two allies, united by the NORAD agreement, integrated their headquarters and operated together, with both CONAD in the United States and the Royal Canadian Air Force Air Defense Command remaining independent commands. The commander in chief of NORAD (CINCNORAD) doubled as commander of CONAD. NORAD's mission was to provide early warning to both Canada and the United States, in the event of a Soviet attack against the United States.[18]

Air Force leaders, most notably Gens. Benjamin Chidlaw and Earle Partridge, guided the planning for ADC programs during the mid-1950s and were largely responsible for how the command operated. The Air Force provided the interceptor aircraft and planned the upgrades needed over the years. The Air Force in conjunction with the Canadian military also developed and operated the extensive early warning radar sites and systems, which acted as "trip wire" against air attack. In addition to the radar sites in Canada, the US Navy element, now Naval Forces NORAD, operated radars and picket ships on both the East and West coasts of North America. The complexity of NORAD's mission was centrally controlled first from an operations center at Ent Air Force Base and later from inside the Cheyenne Mountain Air Force Station complex. In a theoretical scenario, NORAD detected Soviet bombers by one of the early warning lines or picket ships and then ordered interceptors launched, surface-to-air missile sites placed on alert, and SAC alerted to scramble its bombers.

Although the F-86 operated in an inceptor role in the 1950s, the aircraft most recognized for this mission were the F-100 series.

The F-100 to F-106 became known as the Century Series. Of all of these aircraft, the F-105 Thunderchief, more commonly called simply the Thud, went on to have something of a second life in the skies over Vietnam as a fighter-bomber aircraft. None of these aircraft were meant to be air superiority fighters. Instead, their movements were closely controlled by ground-controlled interception (GCI) officers from the time they took off on an intercept, through expenditure of munitions, and until they returned to base. These aircraft came in many forms: North American F-100 Super Sabre (more commonly called the Hun), McDonnell F-101 Voodoo, Convair F-102 Delta Dagger, Lockheed F-104 Starfighter, Republic F-105 Thunderchief, and Convair F-106 Delta Dart. This entire series of aircraft were a mix of fighter-bombers and interceptors. TAC used these aircraft (mainly in Europe) as nuclear delivery vehicles: the F-100, F-101, and F-105. But it was the ADC that used the F-101, F-102, F-104, and F-106 as interceptors to stop Soviet bombers. Designed to take off and be guided by ground control toward Soviet bombers, the interceptors would engage and destroy incoming targets. Interceptors would use either conventional air-to-air missiles or the air-to-air AIR-2 Genie nuclear rocket to take out entire bomber streams.

Of course, the fighters were incapable of intercepting every potential Soviet bomber; exercises proved a 30 percent interception rate was the best that could be hoped for. Therefore, NORAD and CONAD also maintained a heavy integrated air defense system consisting of both Air Force Bomarc missiles (to be fired in advance of the interceptors) and the redesignated Army Air Defense Command's Nike and Zeus surface-to-air missiles. These SAMs surrounded government and military sites throughout the United States. While IADS (integrated air defense system) and GCI are attributed to a Soviet way of defense, the USAF and US Army used both extensively throughout the 1950s and 1960s.[19]

Those That Did Not Survive

The USAF left a number of both fighters and bombers on the cutting room floor of aircraft manufacturers across the country. In the post–World War II era, many aircraft never saw full production, saw their procurement numbers cut drastically, or lived an incredibly shortened life cycle due to advancements in technology or their overall designs not surviving beyond the drawing board phase. These aircraft included fighters and bombers whose design, development, or operational history did not mesh with the Air Force's ongoing conception of warfare.

Technology and aircraft design advanced rapidly in relation to the post–World War II bombers. First there was the B-50 Superfortress, an updated B-29, which was meant to serve as a "stopgap measure" before the next generation of bombers could be developed. The Air Force procured 370 of these bombers meant for use as nuclear delivery vehicles, but the advancement from the B-29 and B-50 to more advanced bombers was not one of linear success.[20]

The USAF gave the B-45 the name Tornado, and perhaps no moniker ever fit an aircraft better than this. The nickname was well earned, not for the aircraft's battlefield performance, but for the damage it wreaked in development and procurement. Although the Air Force eventually accepted 142 of a planned 193 aircraft, the light bomber was beset by developmental problems and the fact that it did not fit nicely inside either SAC or TAC. As initially developed for SAC, it was not designed to carry atomic weapons. SAC therefore viewed the bomber as useless. What was the purpose of a bomber if not to drop atomic weapons? SAC planners handed it off to TAC. The light bomber, unfit to deliver atomic weapons and incapable of being used as a true attack aircraft, languished. It was, in the end, too small for SAC and too big for TAC. Coupled with significant operational problems, the B-45 Tornado sputtered into irrelevance.[21]

An official study of USAF bombers notes that "difficulties encountered by B-45 units . . . posed serious operational problems," and that the aircraft's flaws "varied in importance, but were numerous." These included malfunctions of the Gyrosyn compass and emergency brake, "poorly designed bomb racks," inaccurate indicators, and engines that caught fire during ignition. All of this aside, the simple fact that the design of the aircraft initially precluded it from carrying atomic weapons made it something of a chimera in the postwar Air Force.[22]

Undoubtedly, the most famous "bomber that never was" was the XB-70 Valkyrie. It was designed as a high-speed, high-altitude, penetrating atomic bomber, meant to bomb and be gone before any fighter inceptors could get in a position to fire on it. Its high speed (Mach 3+) guaranteed that no Soviet fighter in service or in production could launch in time to stop it. The Valkyrie seemed viable in the early design phase, but development of Soviet surface-to-air missiles raised questions regarding the validity of the Valkyrie's claimed survivability. Thus, the program effectively ceased in 1961, when President John F. Kennedy called for X planes to be developed and used only as experimental aircraft with a focus on "potential" future aircraft.

In the realm of fighter aircraft, the jet that came closest to meeting the perceived needs of the Air Force but did not survive was the F-108 Rapier. Its history aligns closely with that of the Valkyrie. Unlike its Century Series predecessors, it was going to have two of everything, including two engines and two pilots. The USAF conceived the Rapier as an all-weather, high-speed (Mach 3) interceptor. The Air Force planned to procure nearly 500 of the aircraft for use by CONAD and NORAD, but the entire program was scrapped in 1959. Along with Canada's canceled CF-105 Arrow, the F-108, purpose-built for air defense, never made it into active service.[23]

Developing the Strategic Officer Corps

At the same time that the USAF developed and fielded its post–World War II force structure, it was also changing the way it developed its officer corps from basic college education through its senior leadership. It would be inappropriate to look at the Air Force's conception of warfare without looking at how the organization trained and educated its officer force.

Indoctrination into the Air Force culture began with the obtainment of a bachelor's degree either through Reserve Officer Training Corps (ROTC) or Officer Training School (OTS). Senior leaders of the USAF also wanted their own military academy. The establishment of a separate United States Air Force Academy had long been the dream of senior Air Force leaders dating back to before World War II. In retirement, General Kuter remembered, "Santy Fairchild and I had drawn up a lot of futuristic charts of a separate Air Force, and there always had been an Air Force Academy in it. There always had been an Air Force professional educational system, the concept of the Air University. This dated back to the late 1930s. So my interest in the Air Force Academy was a longstanding one."[24]

Congress created the United States Air Force Academy (USAFA) in 1954. Finalists for the site of the future academy included Alton, Illinois; Lake Geneva, Wisconsin; and Colorado Springs, Colorado. Secretary of the Air Force Harold Talbott announced Colorado Springs as the USAFA's future home in 1954. The Air Force was also granted permission to have members of the graduating classes of West Point and Annapolis switch to service with the Air Force in the early 1950s. The USAFA's first class reported in the summer of 1955 (class of '59). Since the academy grounds were not completed, the new cadets had to report to Lowery Air Force Base in Denver. Nevertheless, like its sister services, the United States Air Force now had its very own academy and a direct commissioning source. Combined with ROTC, which provided the bulk of the Air Force's new officers each year, the Air Force now had a clear path for an officer's development over the course of a twenty-year career.

Officer training beyond the bachelor's degree level continued through learning to fly and employ a weapon's system, or to perform a support function including maintenance, logistics, or personnel management. In addition to initial education and training, the Air Force expanded its own professional military education (PME) schools as well. These included a varied series of educational experiences dependent on rank: Squadron Officer School (SOS) for captains, Air Command and Staff College (ACSC) for majors, and Air War College (AWC) for lieutenant colonels. The Air University at Maxwell Air Force Base designed each course to prepare an officer for the next level of command or, in many cases, the transition to staff work. The latter two programs, ACSC and AWC, for majors and lieutenant colonels, also taught officers how to *think* about warfare on increasingly larger scales at the operational and tactical levels and exposed officers to other service officers and international partners as well.[25]

Conclusion

Once the Second World War was over, the United States Army Air Forces rapidly moved toward independence as a separate service. The raison d'être of this new organization was its ability to deliver of atomic weapons against targets of strategic importance. However, with the focus on strategic air warfare and strategic bombardment as the most probable way of war for the future of the US armed forces as a whole, Strategic Air Command became primus inter pares. The tactical skills of Air Force fighter pilots atrophied.

Strategic bombardment had developed a strong attachment to the popular American conscience. The Air Force very successfully welded its strategic identity to the American people, and "bomb them back to the Stone Age" became part of the American vernacular in the 1950s and 1960s. Strategic bombardment *was* the Air Force. Its entire identity was cloaked in the perception that only the big bombers and the men who flew them high, straight, and true could win the next battle. As late as 1984, the concepts of strategic bombardment could be found even in the work of Dr. Seuss. In his *Butter Battle Book*, an allegory of the arms race, a new aerial invention was seen as the solution to one combatant's problems: "This machine was *so* modern, *so* frightfully new, no one knew quite exactly just *what* it would do!" So deeply embedded was this thinking that it could be dislodged only by a paradigm shift. The Air Force needed an identity crisis if it was going to survive as an effective fighting force, and it needed new innovative thinkers to provide the catalyst for change. Despite claims to the contrary, the USAF never warmly received innovative thinkers in the institution. SAC did not want outside-the-box thinkers. It wanted officers capable of following a prescribed checklist.[26]

Beyond the dominance of Strategic Air Command, the United States Air Force also spent considerable time, money, and labor focusing on the defense of the homeland. Although generations of Air Force officers have been taught that airpower is an inherently offensive weapon, the USAF built an amazing and intricate defensive aerial shield to protect the United States from incoming threats. The stand-up of Air Defense Command followed by the creation of

NORAD in 1957 set the stage for the use of fighters in the role of defensive interceptors. American bombers would be prepared to strike the heart of the Soviet Union, while the interceptors would stop the Soviet bombers from doing the same to the United States. At the same time, the USAF used ROTC, OTS, and the new United States Air Force Academy to train a new officer corps in the 1950s and developed a career-long educational system to develop these officers.

Despite all the planning and posturing for the "inevitable" conflict with the Soviet Union, the diplomatic necessity of containing the spread of communism saw conflict not directly between the two ideological foes, but rather war via proxies on a smaller scale. Instead of a massive confrontation with the Soviet Union in Europe, the next conflicts began in the far-off regions of East Asia.

CHAPTER 5

Strategic Aberrations

Korea and Vietnam

> We went over there and fought the war and eventually burned down every town in North Korea anyway.... Over the period of three years or so, we killed off, what, twenty percent of the population of Korea as direct casualties of war, or from starvation and exposure?
>
> —Gen. Curtis LeMay, 1984

In writing this book and determining where to draw lines and divisions, it became apparent that Korea and Vietnam belonged in the same chapter. Yet, the two conflicts seemed to exist on different planes. While both took place in East Asia and both involved Cold War conflicts in countries where an arbitrary line existed to be sorted out at a later date and time, the similarities ended there. Korea, at least on the ground, looked much like the Second World War. In the air, the B-29 remained the dominant bomber. Vietnam, from a seat in the twenty-first century, looked to be a bridge into a new age of counterguerrilla warfare. Seemingly, the only way the two conflicts could be linked from the perspective of the Air Force was that both clearly were not the type of conflict the Air Force spent the early days of the Cold War preparing for. From the perspective of the leaders of the USAF, all of whom were veterans of the Second World War, neither North Korea nor North Vietnam were the enemy the Air Force prepared to face.

The Korean and Vietnam Wars proved to be aberrations that the Air Force was not prepared to accept. The conflicts did not fit into the Air Force's conception of warfare in the post–World War II environment. The USAF fought these proxy wars in the air using tactical jet aircraft and strategic bombers. In both conflicts, what was strategic and what was tactical use of airpower blurred.

Tactical prowess for air-to-air missions proved to still be a valid combat skill, although it took the Air Force until the end of the conflict in Vietnam to accept this fact. This chapter looks at how tactical fighter pilots demonstrated their ability in both air-to-air and air-to-ground combat and how strategic bomber pilots found themselves struggling to provide a strategic impact, but quite capable of conducting tactical support in the form of close air support. In the end, these two proxy wars demonstrated that a new "style" of warfare had matured and tactical aircraft fighters ascended to have a more prominent role in the USAF.

Korea

The conflict in Korea blended "old and new" in terms of airpower, or what historian John Andreas Olsen called a "transition from the era of the propeller-driven airplane to the era of the turbojet." Aviation historian Kenneth Werrell noted, "Just as American airmen in World War II began the conflict with fighters inferior to the enemy, American airmen in the Korean War found themselves in the same predicament in late 1950." By the end of the war, despite the stalemate on the ground, the USAF and United Nations partner air forces held a decisive edge in the air. Having this advantage, however, did not mean that airpower was the decisive force in Korea despite the USAF's desire to present it that way. The USAF used bombers, namely the B-29 from World War II, and introduced newer jet fighters, the most iconic being the F-86 Sabre.[1]

Still a Strategic-Focused Air Force

The Air Force set about using the tried-and-true method of strategic bombardment against North Korea. Though often overlooked in accounts of the war, the toll the US bombing campaign took against the landscape of North Korea was immense. Gen. Emmett "Rosie" O'Donnell, commander the 15th Air Force of the Strategic Air Command, received orders on 8 July 1950 from Generals Vandenberg and LeMay to head to Korea and direct the strategic bombing campaign against North Korea. Although detailed to the Far East Air Force (FEAF) for the duration of the campaign, all air leaders clearly understood that the B-29 bombers still belonged to SAC. Testifying before Congress upon his return, General O'Donnell adamantly stated on multiple occasions that while he commanded the bombing campaign in Korea, he still belonged to SAC and was simply "on loan" to the FEAF.[2]

O'Donnell had a clear understanding of his role in attacking targets in North Korea and entered the conflict certain of what the bombing campaign could accomplish: "It was my intention and hope . . . that we would be able to get out there and to cash in on our psychological advantage in having gotten into theater and into the war so fast, by putting a severe blow on the North Koreans . . . telling them that they had gone too far in what we all recognized as being an act of aggression." O'Donnell continued, saying he hoped his campaign might go "to work burning five major cities in North Korea to ground."[3]

For bomber pilots, the attacks on North Korea mirrored the industrial raids and firebombings of the Second World War; these were part of the Operations Strangle (I and II) and Saturate. Strangle was an interdiction campaign (US Air Force, US Navy, and Marine air participated) that attempted to stop the flow of goods and supplies from north of the Yalu River. The actual bombing of the supplies where they were most vulnerable—on the Chinese side of the Yalu—remained forbidden. This first operation focused mainly on the roads crossing north to south. Later in 1951, Strangle II attacked traffic moving via the rail system. Operation Saturate, a similar campaign, also targeted segments of the rail lines. None of these campaigns proved overly successful. Airpower historian Conrad Crane noted that initially during the conduct of these operations, "enemy repairs could not keep up with the destruction," but that "enemy countermeasures soon turned the tide. The Communists built duplicate highway bridges across key waterways and cached whole bridge sections near important crossings so repairs could be completed quickly." The Communists' countermeasures did not stop American Air Force officers from reporting that the aerial campaign was producing results.[4]

Gen. Rosie O'Donnell stated before Congress, "I would say that the entire, almost the entire Korean Peninsula is just a terrible mess. Everything is destroyed. There is nothing standing worthy of a name." The bombing campaign against North Korea proved to be a pivotal moment for those that survived it, and its repercussions are still felt in the country today. Many North Korean citizens remember the bombs falling from the sky as the United States Air Force used industrial web theory against a country that was far from industrialized. Of course, North Korea's lack of industrial infrastructure—military supplies came from outside sources—meant the bombing campaign in fact accomplished few meaningful results.[5]

From an air perspective, both sides in the Korean conflict had the benefit of having "sanctuary zones." For the United States, these were the bases in Japan. For the North Koreans, these were the bases located just across the Yalu. When close to the border, American pilots could look down and see MiGs parked on the runways. A similar aggravating scene would present itself again during the Vietnam War as American fighter pilots had to overfly the MiG base at Phúc Yên. Still, the MiGs did not stay exclusively on the Manchurian side of the border, and when they did come up to engage in combat, what resulted was one of the iconic air battles in Air Force history: the dogfights of MiGs versus Sabres.

Sabres and MiGs

Author James Salter, an F-86 pilot during the Korean conflict, said in his novel *The Hunters*, "MiGs were everything. If you had MiGs you were a standard of excellence. The sun shone upon you. The crew chiefs were happy to have you fly their ships. The touring actresses wanted to meet you. You were the center of everything.... If you did not have MiGs, you were nothing." In other words, for the fighter pilot in Korea, having—that is, *killing*—a MiG was everything. It became the ultimate status symbol in an age where Americans back home still gloried in the chivalric stories of the knights of the sky, even if that perception was far from the reality of air-to-air combat.[6]

The MiG-15 was first produced in 1950. Although North Koreans engaged with the Americans, many of the pilots who flew the MiG-15 during the Korean conflict were from China and the USSR. This little, lightweight, easily produced, and perceived technologically inferior aircraft shocked the West. It was, without hyperbole, a better aircraft than anything flown by the United States at the time. It had a more powerful engine than its US counterparts did. At altitude, it could out-turn American aircraft as well.

The American response to the MiG-15 was the introduction of the F-86 Sabre, the sleek single-engine, silver fighter armed with six .50 caliber machine guns. An aura enveloped the Sabre pilots and their legacy. Future astronauts Buzz Aldrin, John Glenn, and Gus Grissom all flew the F-86 in Korea, as did NASA flight director Gene Kranz. The fights involving the MiG-15 and F-86 became emblematic for the Air Force during the course of the conflict. Whereas after Vietnam, it would become standard for pilots to train against dissimilar aircraft, since by that point, Eastern and Western aircraft had diverged significantly in shape, size, and aerodynamic principles, this was not true in the skies over Korea. In many ways, the F-86 and MiG-15 were similar. An untrained eye might not be able to differentiate between the two. Both had a single-engine intake at the nose of the aircraft and swept wings along the fuselage. Only any distinct coloration, unit organization markings, and the horizontal stabilizers at the back of the aircraft (on the tail of the MiG-15 and lower on the F-86) were clear indications of a difference.

Performance-wise, the MiG-15 maintained a higher ceiling, which was a serious disadvantage for the American pilots, who often had to wait for the MiGs to initiate engagements from their higher altitude. If the MiGs wanted to remain above the American aircraft, there was little to nothing American pilots could do except stare at the contrails of their enemy so tantalizingly close.

In order to be in a position for an engagement with a MiG, American pilots first had to fly the two hundred miles north to the area known as "MiG Alley." From that point, US pilots had to hope that the enemy MiGs, flying from sanctuaries in Manchuria, felt like engaging that day. Werrell stated, "MiG Alley is best remembered for the glory of fighter pilots engaged in the first jet-versus-jet aerial combat."[7]

When the MiG-15s did decide to engage in air-to-air combat, they did have one significant disability: their pilots. Historian Conrad Crane stated, "In the key areas of experience and aggressiveness, American pilots had a big edge." American pilots would again possess such an advantage during Desert Storm and Allied Force. A North Korean fighter pilot who escaped to South Korea in 1953 also said of his aircraft, "The MiG-15 was good, but hardly the superfighter that should strike terror in the heart of the West. There was no question that the F-86 was the better fighter."[8]

That said, every statement above certainly reflects a Western perspective of the Korean conflict. Scholarship by historian Xiaoming Zhang presents a far different picture. Chinese sources state that the "PLAAF . . . engaged in 366 battles, claiming to have shot down 211 F-86s, 72 F-84 and F-80 fighter-bombers, and 47 other types of planes. . . . They acknowledged the loss of only 224 MiG-15s, 3 La-11s, and 4 Tu-2 bombers (totaling 231)." By way of comparison, the aerial victory tables for the United States Air Force held at the Air Force Historical Research Agency have over 900 full or partial victories credited to the USAF. This does not include any Navy or USMC victories. Likely there will never be an agreed-upon accounting of claims and losses by either side. Jon Halliday called this "an insurmountable methodological problem" as there is not an internationally accepted practice for what counts as an aerial victory.[9]

Conclusion

The Air Force way of war in Korea was not entirely different from what it had been during the First and Second World Wars. What mattered to fighter pilots was still being a "good stick," closing with the enemy—if possible from an unannounced six o'clock position—and gunning him into the ground. The attainment of great fighter pilot status was still measured by the number of kills. Over thirty Air Force pilots reached the designation of ace during the Korean War—names still discussed in certain circles: "Boots" Blesse, "Gabby" Gabreski, James Jabara, Joseph McConnell, and "Robbie" Risner. The point of being a good fighter pilot was gaining air superiority, if only temporarily, so that the bombers could accomplish their mission without prohibitive interference from the enemy. For bomber crews, the mission remained the systematic destruction of industrial targets in order to shorten the war and force the opposing side to concede defeat or, at least, sign an armistice. Although the USAF's aerial war over Korea is best remembered as the contest between MiGs and Sabres, bombardment doctrine reigned supreme.[10]

Air Force Chief of Staff Gen. Hoyt S. Vandenberg proclaimed in 1950 that "the Air Force is on trial in Korea." Writing for the prosecution, USAF Col. James T. Stewart in his book *Airpower: The Decisive Force in Korea* stated unequivocally, "Without question, the decisive force in the Korean War was airpower. Through its unrelenting efforts in those dark days of the summer of 1950, US and other U.N. ground forces were able to retain a foothold on the peninsula. During the three years of fighting that followed, defeat or victory often depended upon the successful accomplishment by the United States Far East Air Forces of the tasks laid upon them. Never once during the struggle did our air forces relinquish the offensive to the communists."[11]

For the defense, in a more critical view, Alan Stephens noted, "Notwithstanding the drama of the combat above the Yalu, the air war in itself was unlikely to realize the rapid strategic effect that has been one of the constants of air power theory.... The campaign did not put any pressure on either China or the Soviet Union.... It remains a moot point whether strikes against first-world targets such as electricity and infrastructure can generate a strategic effect in third-world countries."[12]

If Vandenberg was correct in calling it a trial, then the verdict was a hung jury. The problem was that strategic bombardment against industrial targets was difficult when the number of industrial targets was limited.

Over the three-year conflict (that has stretched into a nearly seventy-year stalemate), the United States Air Force—along with the aerial components of the Navy and Marine Corps—gained air superiority, lost it with the entrance of the MiG-15s, and regained it with the entrance of the F-86s. In Korea, the Air Force aided in destroying the industrial resources in a nation that was largely nonindustrial but encountered limitations in applying strategic bombing theory against a country and targets whose war-making capabilities could not be touched, since those were located in China. Helicopters were used for the first time in search and rescue operations, a mission type that became a mainstay of the Air Force's special operations community.

However, no sooner had both sides signed an armistice than the tactical arm of the Air Force allowed itself to forget the aberration of Korea and leave it behind. As Conrad Crane so eloquently explained, TAC now "struck a Faustian bargain with the atomic Mephistopheles, transforming the organization into a 'junior SAC' concentrating on the delivery of small nuclear weapons," and "the Air Force, and TAC with it, soon returned to its focus on general nuclear war." Fifteen years later, payment on this bargain was called due in the country of Vietnam.[13]

Vietnam

A slow and arduous slippery slope brought American airpower, and the US Air Force, into the conflict in Vietnam. Although generally viewed as a single overarching conflict, in terms of airpower, the Vietnam War is best seen as multiple air wars: an in-country war, which provided close air support to troops on the ground in South Vietnam; an out-country air war in North Vietnam, both strategic and tactical; and an interdiction campaign waged largely in Laos and Cambodia along the Ho Chi Minh Trail. The USAF entered into South Vietnam ostensibly to aid in the training and equipping of South Vietnam's Air Force. This changed after the Gulf of Tonkin incident when large numbers of US troops and aircraft funneled into the region to both aid the South in its fight against the Viet Cong and to make war against North Vietnam. At this point, the USAF prepared to use nearly every aircraft in its inventory—from propeller-driven O-1 Bird Dogs, to fighters and fighter-bombers of every conceivable type, to light, medium, and heavy bombers, including the B-52s—to prosecute the war in Vietnam and the surrounding countries.

Command and Control and Route Packages

As noted by historians who have delved into command and control (C2) of air assets during the Vietnam War, the C2 arrangements throughout the conflict were nothing short of a complete mess. The 2nd Advanced Echelon (ADVON); later, the 2nd Air Division; and finally, the 7th Air Force led efforts of USAF assets inside of South Vietnam and Thailand. The latter country was where many assets were stationed, at Royal Thailand air bases, not including strategic assets (B-52s), which were retained by SAC. SAC did provide a liaison officer to 7th Air Force headquarters. The 7th Air Force also did not control the Navy air assets working off the Vietnamese coast (except when those assets attacked targets in South Vietnam), USMC air assets, or Army rotary-wing assets, each of which fell under the command of separate C2 organizations. The commander of the 7th Air Force reported directly to the commander of the Pacific Air Forces (PACAF), while also serving as the air deputy to the commander of Military Assistance Command–Vietnam (MACV). The commander of 7th controlled Air Force assets attacking targets in North Vietnam but was never able to gain command and control (or even operational control) of Navy assets in the same area. The commander in chief Pacific Fleet (CINCPACFLT) successfully argued that "naval Airpower was an inherent part of the fleet, and its mission and could not be separated."[14]

To keep these lines of separation clear, the United States Pacific Command divided North Vietnam into route packages (RPs). These were RP-1–5, RP-6A, and RP-6B. The USAF controlled RP-5 and RP-6A; the US Navy controlled RP-2, -3, -4, and RP-6B. Route package 1, the southernmost RP, belonged to the Navy until 1966 when it transferred to the USAF and MACV. At the highest level, a Navy four-star admiral (CINCPACFLT) controlled air assets in four of the route packages, and an Army four-star general (MACV) controlled air assets in three of the route packages. To make matters worse, the line dividing 6A and 6B fell directly through the capital city of Hanoi. If nothing else can be said of the air wars during the Vietnam conflict, let it be stated that the command and control apparatus was not ideal.[15]

Gen. William Momyer, commander of 7th Air Force, was not pleased with the division of airpower. Momyer commanded 7th Air Force from April 1966 until July 1968. He later wrote, "Dividing North Vietnam into route packages compartmentalized our airpower and reduced its capabilities." From Momyer's perspective, the Navy's Task Force 77 (TF-77) had an insufficient number of aircraft operating off of carriers in Yankee Station (a fixed-coordinate location off the coast of Vietnam) to cover all of the targets in its assigned route packages. Command and control in Vietnam became all about determining how many angels could fit on the head of a particular service pin rather than the importance of employing said angels in a cohesive and unifying manner.[16]

The In-Country War

For the USAF, the war in South Vietnam was about close air support to troops on the ground, what historian John Schlight called the "largest, most sustained ground support campaign in the history of aerial warfare." In the historiography of Vietnam, especially from airpower scholars, this aspect has received less attention, save for familiar place-names like Khe Sanh or Ia Drang, than the bombing campaigns and fighters dueling MiGs that both occurred in the North. In his edited work *A History of Air Warfare*, John Andreas Olsen devotes a chapter to the air war over North Vietnam but does not give the same attention to the war in the South.[17]

Many Americans might still be able to recall Khe Sanh and Ia Drang as significant operations. Literature and film representations have certainly aided in raising the profile of the latter. Other names have slid from memory, including Operation Cedar Falls, America's largest ground operation of the war, and Operation Junction City. In both of these operations and dozens of others, the USAF used its fighters and bombers to support troops on the ground. In situations where B-52s dropped bombs as part of preplanned strikes, the results could be especially devastating.

Some histories, including Schlight's *War in South Vietnam*, separate out B-52 strikes from other fighter strikes during an operation, but the result was really the same. When it came to supporting ground troops in South Vietnam, it did not matter if an aircraft was "tactical" or "strategic." Separating aircraft along such lines became irrelevant so long as strikes placed large amounts of iron on the enemy. Even pilots flying medium bombers like the B-57 jokingly called the F-100 a "minibomber." SAC balked at having B-52s stationed permanently in Thailand to support the war in the South, stating that it harmed their nuclear response and deterrence responsibilities. SAC leaders perceived that every B-52 at U-Tapao Royal Thai Navy Airfield was one not capable of responding to the Soviet Union, but General Westmoreland argued their presence helped the overall war effort. Westmoreland won.[18]

Although the USAF provided substantial support to soldiers and marines on the ground, the ability to quantify what the Air Force accomplished in South Vietnam has remained elusive. Schlight noted that "analysis of the Air Force's effectiveness was extraordinarily difficult"; furthermore, he concluded, "the Air Force possessed neither its own war objective nor enough reliable data to quantify the results."[19]

Air Force support to ground forces and the overall war effort increased over time. When Richard Nixon became president and began inching toward a US withdrawal, airpower took on increasing importance. As troop strength was drawn down, the need for airpower to buttress South Vietnam's forces and provide maneuvering room only increased. In the end, the Air Force failed in a substantial way to accomplish what it entered the conflict initially to do: train and equip the Air Force of South Vietnam (VNAF). The VNAF was not prepared to fend off the North's final attack in 1975. In one Air Force history, Bernard Nalty ends his study of the war in the South by saying,

> The process of Vietnamization, which the Air Force had helped shield, produced size rather than skill. The South Vietnamese air arm, for example, had too few trained men to operate and maintain all its equipment, despite its vast expansion. Moreover, the post-cease-fire training effort could not make up for the deficiencies left by the wartime program. In addition, the inventory of aircraft had huge gaps: the fighter-bombers and attack planes lacked the range to hit targets outside South Vietnam; the transports and helicopters could not support extended operations on the ground; and no bomber could approach the B-52 in destructiveness. Finally, stocks of fuel and munitions declined as American military assistance diminished.[20]

The end game for South Vietnam came in 1975.

The Out-Country War, the 94-Target List, and the Strategic Bombing of North Vietnam

Three distinct campaigns composed the bulk of the air war in North Vietnam: Rolling Thunder (1965–68), Linebacker I (May–June 1972), and Linebacker II (December 1972). The first, Operation Rolling Thunder, used primarily, but not exclusively, tactical fighter aircraft. History remembers the Linebacker operations for their use and losses of B-52 bombers; however, all three bombing campaigns used nearly every available aircraft in the USAF's arsenal. Looking at these operations in turn demonstrates that airpower employed against North Vietnam, as noted by such historians as Mark Clodfelter, Jacob Van Staaveren, and Wayne Thompson, was a "tale of air power badly used."[21]

An article in the USAF's *Air and Space Power Journal* in 2001 noted that contrary to popular belief that the Joint Chiefs of Staff (JCS) wanted to bomb Vietnam "back to the Stone Age," as Gen. Curtis LeMay famously stated, by using the traditional industrial web theory, the planning process for Rolling Thunder was actually quite perceptive as far as air campaign planning went. The 94-target list that planners came up with did not focus on industrial targets *primarily*. Rather, airfields, lines of communications, barracks, headquarters, and supply dumps were to be hit first and hardest prior to traditional industrial targets, which included chemical plants, power plants, machine tool factories, and iron and steel plants. The target sets—divided into three categories, A, B, and C—were of a much more tactical and operational nature than strategic.[22]

This change of perception on what the JCS wanted to strike is all well and good, and the debate about whether "more bombing earlier" would have made a significant difference in the outcome of US involvement in Vietnam will not be solved in these pages. The simple fact remains that President Lyndon Johnson's administration chose not to strike the 94 targets as part of a massive and instantaneous aerial campaign but elected to slowly apply pressure against North Vietnam through an ever-increasing bombing campaign meant to eventually ratchet up the weight placed against the North to an unacceptable level. However, it seems apparent that no amount of aerial pressure was going to force the North Vietnamese into negotiations in the 1960s; the course of the war was not favoring America's policies or aims.

Rolling Thunder

If historians can agree on anything, it is that Operation Rolling Thunder was an abysmal failure, at least insofar as an operation that attempted to influence popular and political opinion inside North Vietnam was concerned. The stated objective of persuading North Vietnam to stop aiding the Viet Cong (VC) in the South and, at the same time, of bolstering South Vietnam's position was ephemeral at best. Even the goal of limiting North Vietnam's ability to aid VC fighters in South Vietnam proved nearly impossible. Not only that, but American pilots were forced to fly into the teeth of one of the most technologically advanced air defense systems in the world, composed of layered surface-to-air missiles, antiaircraft artillery (AAA), and MiG fighter aircraft. Finally, the desire of the Johnson administration to keep the war from widening into one against China or the Soviet Union seriously limited the ability of airpower to strike at useful targets. The Johnson administration feared killing Soviet or Chinese advisers at specific targets. Therefore, targets within a thirty-mile ring around Hanoi and a ten-mile ring around Haiphong were off-limits. Targets within a buffer zone along the Chinese border were also off-limits.[23]

There were also concerns inside the administration that if a combined air offensive were to strike all of the industrial targets inside North Vietnam, including those inside Hanoi and Haiphong, the North Vietnamese still would not agree to American political goals. There was also a concern that striking all possible targets on the 94-target list could mean that American airpower would run out of targets to strike. It was a maddening circular logic. Therefore, inside the Oval Office, President Johnson and a cabal of advisers selected targets on an ad hoc basis. Historian Earl Tilford noted, "The targeting bore little resemblance to reality in that the sequence of attacks was uncoordinated and the targets were approved randomly—even illogically. A bridge might be struck one day and a radar site the next. The targets most coveted by the Air Force, the factories . . . and the power plants, were also off-limits." Some of the targeting for Operation Rolling Thunder conducted in this ad hoc manner occurred at the Tuesday luncheons held at the White House and attended by President Johnson, Secretary of State Dean Rusk, Secretary of Defense Robert McNamara, National Security Advisor Walt Rostow, and Chairman of the Joint Chiefs Gen. Earle Wheeler, among a list of others who came and went week by week. This, then, was the heart of the Rolling Thunder operation: an ad hoc air war against targets chosen by committee and not air planners with missions flown into the teeth of one of the most densely populated IADS on Earth. It was a recipe for disaster.[24]

Much has been made of the gradual nature of the bombing campaign in the North with many, particularly those in the Air Force, asserting that a more sustained campaign against the North earlier in the war could have a forced a settlement sooner. Such an outcome was highly unlikely. The context of the war in Vietnam, both in the North and in the South, was different in the latter part of the 1960s than it was later in the war—that is, the Johnson administration held real concerns for the war expanding due to the intervention of China or the Soviet Union. Nixon's foreign policy would successfully alleviate some of these concerns. It seems highly unlikely that even a sustained and decisive aerial campaign anytime in the 1960s could have forced the North Vietnamese to reach a different conclusion on their desired end state. If Rolling Thunder had been an immediate or "instant" thunder, the damage might have been greater and lives lost might have higher, but the outcome would have remained largely the same. Between the cessation of bombing against North Vietnam in 1968 and the suspension of Rolling Thunder, the bombing did not actually cease, but rather was shifted to targets in Laos and Cambodia.

The air-to-ground war against the supply routes that ran from North Vietnam through Laos and Cambodia and into the South remained largely unknown to the American people. American pilots gave it the name Ho Chi Minh Trail; to those in the North, it was the Truong Son Strategic Supply Route. It ran from two primary entrance points in the North—the Mu Gia Pass and the Ban Karal Pass—traveled south through Laos and into Cambodia. At numerous exit points, men and materials were dropped off in South Vietnam. The US attacks on the trail included operations in Laos codenamed Barrel Roll and in Cambodia, Steel Tiger. The USAF spent the entirety of the conflict in Vietnam interdicting the trail, with various levels of success.

Linebacker I and Linebacker II

The situation and context against which the air war was prosecuted had changed by 1972 and the beginning Operation Linebacker. For starters, President Nixon had already announced his intention to remove American troops from South Vietnam. He had also used diplomatic pressure. His "détente" with China and the thawing of relations with Moscow put pressure on North Vietnam to end the conflict.

North Vietnam planned to launch an all-out offensive in the spring of 1972, and Nixon responded in kind with a massive interdiction campaign against all targets in the North. Nothing was off-limits: bridges, industrial targets, the air defense system, and MiG airfields. Another reason that Linebacker differed significantly from earlier uses of airpower was the introduction of the first precision-guided munitions for the air-to-ground portion and better-equipped and -trained pilots to fight the air-to-air war—particularly for the US Navy. The capabilities of the American military had changed significantly going into the final months of the conflict, something not possible just a few years earlier. The North's "Easter Offensive" began on 30 March 1972, and Linebacker I no doubt blunted North Vietnam's offensive through its massive interdiction campaign and attacks against the North's industrial targets in and around

Hanoi and Haiphong. Linebacker I probably hastened an end to an already drawn-out war. With improved technology, a willingness to allow airpower more targeting options, and different contextual end states for both the United States and North Vietnam meant that what airpower was capable of providing as a tool by 1972 was fundamentally different from what it could provide earlier in the war. Nixon benefited from this. However, Linebacker I was not the final move in this conflict. Linebacker II proved to be the endgame, and the outcome of this final campaign depends entirely on which side you ask.

Was Operation Linebacker II a success? The answer generally depends on whom you ask and where they are from. The USAF clearly believes it was a victory for the United States and the USAF and that it helped end the war. The National Museum of the United States Air Force purports that "B-52 aircrews dropped over 15,000 tons of bombs on important military targets during Linebacker II and helped force the North Vietnamese back to the peace table." The Air Force's Historical Studies Office argues: "By 29 December 1972, the 700 nighttime sorties flown by B-52s and 650 daytime strikes by fighter and attack aircraft persuaded the North Vietnamese government to return to the conference table." However, both official sites admit that the Air Force paid a high price for this "victory," including the loss of fifteen B-52s and twelve fighters between the Air Force and the Navy.[25]

Linebacker II followed several months after Linebacker I had broken North Vietnam's Easter Offensive. In the peace negotiations an agreement had been reached between North Vietnam and the United States, but the South balked. Therefore, not only to force a settlement to the conflict with the North but to prove to South Vietnam that the United States could protect the country in the wake of troop withdrawals, Nixon authorized the bombardment of northern cities, including Haiphong and Hanoi.

The first nights of attacks followed a predictable pattern, with American B-52s approaching Hanoi from the north and then turning south to make their bombing runs against targets on the outskirts of the city. This meant the North Vietnamese forces needed only to wait for the bombers to appear over or near their targets to send up a barrage of SAMs. Furthermore, as the B-52s conducted their post bomb run turn, their electronic countermeasure gear, which pointed down during flight, angled away from the ground and exposed the aircraft at their most critical and vulnerable moment.[26]

Eleven B-52s were lost in the first four nights of attacks, and four more were shot down over the coming days. SAC and 8th Air Force, after being strenuously lobbied by the B-52 commander at U-Tapao Royal Thai Navy Airfield, Brig. Gen. Glen Sullivan, changed the approach vectors, and as a result, losses decreased. North Vietnam was also running short of SA-2 missiles. A peace treaty was signed by all parties a month later. Historian Mark Clodfelter noted that "the conviction that air power played the decisive role in gaining an agreement permeated the Air Force."[27]

Both the United States and North Vietnam claimed victory in this final conflict. While the Air Force has a rich history of believing Linebacker II helped end the war, the North Vietnamese view the very same event as a victory for their SA-2 missile crews. In North Vietnam the battle is known as "Dien Bien Phu in the skies." Historian Merle L. Pribbenow translated *Victory in Vietnam: The Official History of the People's Army of Vietnam, 1954–1975*. This account states that "at 7:00 A.M. on 30 December 1972, the US government was forced to announce a bombing halt north of the 20th parallel and suggested that a meeting be held with representatives of our Government in Paris to discuss signing a peace treaty. The enemy's massive strategic offensive using B-52s against Hanoi and Haiphong had been crushed. Nixon's dream of negotiating from a position of strength had ended in total failure."[28]

While the People's Army of Vietnam's official history certainly emits the scent of propaganda, it remains instructive to note that both sides claimed a significant victory that forced the other to sign a peace treaty. If nothing else, Linebacker II did secure the release of American pilots held as prisoners of war, and for that it was a success.

In the end, North Vietnam considered the blunting of the American bombing effort and the downing of the fifteen B-52s a substantial victory. North Vietnam built museums around the downed American bombers. Does this mean that the American perspective that Linebacker II won the war is based on a lie? It is really just a matter of perspective. The United States and North Vietnam saw in the conduct of operations in December what they wanted to see. For North Vietnam, it was a refutation of American airpower. For the United States, it represented achievement of an honorable exit from a war that had dragged on for too long and cost too many lives.

Fighters and Fighter-Bombers

Rolling Thunder, the Linebacker operations, plus a number of other named operations over North Vietnam, Laos, and Cambodia accounted for the bulk of USAF campaigns during the Vietnam War. At the same time, senior Air Force leaders continued to scrutinize closely the conduct of the air-to-air and air-to-ground war waged by fighters and fighter-bombers. As the primary function of an air force is to gain and maintain control of the skies so that other assets can prosecute the air-to-ground war, the war in Vietnam found the USAF's fundamental purpose seriously challenged throughout the conflict. While the Air Force also flew F-100 Super Sabres, F-104 Starfighters, A-7 Corsairs, and later F-111 Aardvarks during the Vietnam conflict, to gain the best picture of "fighter" operations we will focus on the two principal fighters that engaged in combat, the F-105 and F-4.

The Air Force's F-105 Thunderchief (the Thud) became the Air Force's air-to-ground workhorse during the Vietnam War. Initially designed in the early 1950s as a high-speed nuclear delivery vehicle with a limited air-to-air capability, the F-105 morphed into a fighter-bomber. As the Vietnam conflict escalated, the F-105D saw continuous changes made to its design, including the addition of armor plating and alteration of bombsights (a change from the bombsight designed for nuclear delivery). The F-105's payload was immense; it commonly carrying twelve 750-pound bombs. Even at its fully loaded weight, it remained an extremely fast aircraft. The Air Force lost 334 F-105s during the war, the vast majority to AAA, owing to the requirement to work at lower altitudes. These losses accounted for 40 percent of the Air Force's F-105 inventory.[29]

Perhaps more emblematic of the air war was the F-4, which served in both air-to-air and air-to-ground roles during the war. Designed for the Navy, it was selected (some might argue ordered) by the Air Force in 1962. The F-4 accounted for 108 of the 137 air-to-air kills made by the Air Force during the Vietnam War. The most successful weapon employed by the F-4 fighter pilots and weapons officers was the AIM-7 Sparrow, followed by the AIM-9 Sidewinder. The USAF produced three aces during the war: Capt. Charles "Chuck" DeBellevue, Capt. Steve Ritchie, and Capt. Jeff Feinstein. Of the three, Ritchie was the only pilot; both DeBellevue and Feinstein were weapons system officers. Four of the kills awarded to Ritchie and DeBellevue came when flying together.[30]

A paradigm shift occurred during the Vietnam War. Since the heady days of the Air Corps Tactical School and strategic bombardment during the Second World War, the bomber pilot and the adherents of strategic bombardment theory had ruled the Air Force. In the 1950s and 1960s, Strategic Air Command dominated the Air Force as the primus inter pares. What the Air Force observed during the Vietnam War forced a radical change on the organization. The shift occurred, slowly at first, but gaining momentum throughout the 1970s—a discernible move away from the strategic toward the tactical.

This sea change entailed a massive training revolution, which occurred after the war, leading to the ascendancy of tactical fighter pilots throughout the senior ranks of Air Force leadership. The "rise of the fighter generals" and Tactical Air Command culminated in the selection of Gen. Charles Gabriel, a fighter pilot, as Air Force chief of staff in 1982. Since that date, no bomber pilot has held the top job in the USAF. The tactical arm of the Air Force grabbed the reins of leadership and has never let go. It was a paradigm shift in how the Air force planned for and executed combat.[31]

Conclusion

The conflicts in Korea and Vietnam both proved to be antitheses to the US Air Force's preferred way of war, but both conflicts also profoundly shaped USAF thinking about future organization, force structure, and engagements. The USAF took many valuable lessons from these two conflicts, but primarily from Vietnam. Junior officers swore they would never allow the mistakes made by more senior leaders to ever be repeated—at least what they perceived to be mistakes by senior leaders. The Air Force spent a fair amount of time in writing official and quasi-official histories of the conflict. The Air Force's History and Museums Program produced no fewer than twenty-five books, studies, or monographs dealing with issues of the Vietnam conflict, covering everything from interdiction along the Ho Chi Minh Trail to civil support in South Vietnam. This does not include other reports and studies conducted by various Air Force organizations such as the Red Baron Reports, which profoundly shaped both exercises and doctrine in future operations, and the Contemporary Historical Evaluation of Combat Operations (Project CHECO) reports. The CHECO reports comprise 250 studies on various aerial operations during the Vietnam War.[32]

All of these reports, studies, books, and monographs helped influence the paradigmatic shift that occurred after the Vietnam War ended. Perhaps the best that could be said of Vietnam for both the Navy and the Air Force was that it proved to be a learning experience from the top down. Senior leaders and junior pilots alike came out of Southeast Asia prepared to make changes to force structure, technological advancements and incorporation, and training environments and programs. The post-Vietnam era found the Air Force prepared to alter its preferred method of warfare. The age of strategic dominance was at an end, and the ascension of tactical airpower was readily apparent. When combined with technological advancements in every facet of procurement from weapons to aircraft, the USAF changed dramatically after Vietnam.

EPOCH III

Tactical Ascendancy, 1975–2019

The F-15 was a twin-engine, twin-tailed, air-superiority fighter. At sixty-three feet long and forty-two feet from wingtip to wingtip, this fighter was as big as a tennis court. With two powerful Pratt & Whitney engines, it could reach speeds 2.5 times the speed of sound. Its ceiling was an amazing 65,000 feet. Initially with an AN/APG-63 radar and later with an updated active electronically scanned array (AESA) radar, the F-15 had the ability to detect enemy aircraft at great distances. Capable of carrying a combination of AIM-7, AIM-9, or AIM-120 air-to-air missiles as well as an M61A1 rotary cannon, the F-15 could reach out and kill beyond visual range or perform in a close-in turning dogfight and kill with equal success. It was a pure fighter aircraft.

Then there was the F-16: a sleek, fast, nimble and comparatively small fighter. The F-16 became a jack-of-all-trades. Air-to-air, air-to-ground, suppression of enemy air defenses (SEAD), close air support (CAS)—the F-16 did it all. A favorite song among F-16 pilots goes, "We've got every mission that you do and we fly 'em all better than you . . . we're single seat, multi-role."[1]

These two aircraft became the backbone of the USAF's combat air force (CAF) in the 1980s, and both proved their worth in the engagements of the 1990s. Different variants of the aircraft conducted both air-to-air and air-to-ground missions. The men—and for the first time women—who flew these aircraft into combat trained differently than their predecessors. All of them were college educated. Early on, they were exposed to John Boyd's energy maneuverability and fast transient theories. They sharpened their skills at realistic training exercises and deployed around the globe to train with coalition partners. At Combat Archer and Combat Hammer, specially designed exercises, they learned what it felt like to have live missiles and bombs leave the rails. In short, they became masters of the tactical battlefield. Long before engaging in actual combat, their training equipped them with enough realistic scenarios that even if they were not "used to" combat, they were well prepared for it. That the Air Force's "way of war" fundamentally changed in the post-Vietnam era is an accepted fact in the early twenty-first century, and this evolution has been documented by academics, military leaders, and popular authors alike.[2]

The Air Force's way of war after Vietnam was defined by more tactically focused organization and a significantly larger role for tactical aircraft: F-15s, F-16s, F-111s, and F-117s. The bombers were not gone, nor would they ever be, but the Air Force was now led by pilots who had experience flying fighters over Vietnam, Laos, and Cambodia.

CHAPTER 6

A Useful Way of War?

American Airpower after Vietnam

> We were emerging from the problems of Vietnam.... We were losing Air Force people to the airlines at a greater rate than we wanted.... We were trying to expand the force ... [and] maintain it and accept into it new and very difficult equipment. So it was a period of continual but relative crisis.
>
> —*Gen. Robert J. Dixon, TAC commander, 1982*

The years 1975–91 proved to be another "interwar" period for the United States Air Force, situated in the post–strategically focused interim before the rise of significant conflicts in Southwest and Central Asia. Although the perceived major threat remained generalized war against the Soviet Union—in all likelihood to occur in Western Europe—substantial changes in *how* to approach war vis-à-vis that particular threat took place. First, the post-Vietnam Air Force saw the rise of fighter pilots to positions of leadership throughout the service, and the role of the strategic bomber diminished as tactical aircraft moved to the forefront of the Air Force's way of war. As tactical generals rose to leadership positions and Strategic Air Command generals faded into the background, the way in which senior leaders envisioned war happening also changed. Tactical fighters took on much more responsibility. Smaller conflicts in Grenada and Panama demonstrated little, but El Dorado Canyon in Libya proved that tactical fighter-bombers were capable of strategic-level effects. Second, the USAF also spent considerable time training for war, something that would prove its utility throughout the 1990s. The contributions of John Boyd and John Warden were of principal importance to the training and doctrinal changes occurring after Vietnam. Although the phrase "revolution in military affairs" is often thrown around, the period between the end of Vietnam and

Operation Desert Storm proved both revolutionary and evolutionary as new aircraft came on line and those new aircraft were incorporated into the Air Force's overall force structure.

Throughout the 1980s and into the early 1990s, all indicators pointed to the primacy of tactical airpower whether used in a strategic or a tactical manner. Urgent Fury, El Dorado Canyon, and Just Cause all indicated tactical aircraft were now the primus inter pares in the USAF. The necessity of a separate Strategic Air Command was clearly diminishing. This epoch of tactical ascendancy also represented the perfect combination of new technologies that melded with the improved human element to alter the American Air Force's concept of operations and the way it prosecuted the conflicts of the 1990s and beyond.

Changes in the Military Airlift Command and Tactical Air Command

An often overlooked harbinger of the rise to eminence of tactical aviation was the selection in 1982 of Gen. Charles A. Gabriel to be the eleventh chief of staff of the USAF. Gabriel, a veteran of the Korean conflict, was the first pure "fighter" pilot to hold the post. In Korea, he flew F-51s and F-86s, earning the coveted moniker of "MiG Killer" for his two credited aerial victories. Gabriel's tenure switched the traditional assignment of the role of Air Force chief of staff from a member of the "bomber mafia" to a "fighter general." This trend lasted until Gen. Norton Schwartz (an Air Force Special Forces pilot) replaced the fired T. Michael Moseley in 2018. After Schwartz, however, the post again reverted exclusively to fighter pilots.[1]

Although fighter pilots now ran the Air Force from the early 1980s, massive developments also occurred in the area of air mobility, despite being overshadowed by the "pointy-nosed" fighters. Many in the community of cargo pilots might even make a valid argument that their profession should be the true representation of what airpower can accomplish or that they truly make all other forms of aviation possible. Historian Robert Owen noted: "Global Air Mobility is an American invention. . . . Global air mobility changed the world. Most obviously, it elevated the American military's penchant for speed and maneuverability to an unequalled art. Since World War II, it is fair to say that every major US strategic concept and regional plan has presumed substantial reliance on air mobility. Whether contemplating a bomber campaign against the Soviet Union or halting another North Korean surprise attack, American war planners depended on transport and tanker aircraft to launch, reinforce, and sustain operations."[2]

Beginning in 1965 with the introduction of the C-141 Starlifter, the Air Force procured a series of large and technologically advanced cargo and refueling aircraft that allowed fighter aircraft to function around the globe. The C-141, long and cigar-shaped, was the first of the "intertheater" cargo aircraft, built purposefully to fulfill air cargo (both troop and material movement) missions on a global scale. The USAF also considered it another *strategic* asset, placing it alongside the bombers in order of importance. The Starlifter became operational in 1965 and saw service through much of the Cold War, including in both Vietnam and Desert Storm.

It was followed by the largest cargo aircraft ever built by the United States military: the C-5 Galaxy (often called the C-5 FRED [fucking ridiculous engineering design]), designed as a strategic airlifter, but also to carry outsize cargo that would otherwise have to be transported by ship. Between 1968 and 1982, the C-5 held the distinction of being "the world's largest and heaviest aircraft." Appearing operationally in the 1970s, it served during Operation Nickel Grass, Desert Storm, and other missions. In between conflicts, it proved its worth flying outsize cargo and material all over the globe.[3]

In 1973, beginning on 6 October (the Jewish holy day Yom Kippur), the Middle Eastern nations of Syria and Egypt launched an attack against Israel. The surprise attack was initially successful, breaking through Israeli defensive lines and driving the Israeli army back. In what became known as the Yom Kippur War, the Israeli Ground Forces were eventually able to halt the advance. The conflict turned into a stalemate of attrition. The Israeli government turned to the United States and requested munitions and material to replace losses suffered in the opening phases of the war. President Nixon, somewhat reluctantly, agreed to the resupply. Although logistically it is easier to move mass via shipping, time was of the essence. Thus, the task fell to the commander of Military Airlift Command, Gen. Paul K. Carlton, who directed Operation Nickel Grass.

Operation Nickel Grass demonstrated strategic and global airlift in its purest form. With no time for overseas shipping and the need to replace Israeli losses with "outsize cargo" consisting of tanks and other vehicles (in addition to munitions and other supplies), the answer was clearly the C-141 and C-5 strategic intertheater airlifters. Between 14 October and 14 November, MAC moved 22,300 tons of cargo from stateside locations to Israel on 567 flights. While the operation was not as large, as long, or as extensive as the Berlin Airlift of 1948–49, it did prove, on a grander global scale, the reach of intertheater airlift in a crisis situation. The large airlifters often take a backseat to their pointy-nosed brethren in written works, but the importance of air mobility operations and their impact on combat execution and support of allies for the USAF cannot be overstated. From Berlin to Nickel Grass and later Operation Desert Shield, it was air mobility—cargo hauling and air-to-air refueling—that proved to be the linchpin of projecting airpower across the globe. Although de Seversky could not have recognized this when he published his seminal work in 1942, it was air mobility that was the key to allowing "victory through air power."[4]

The newest member of the cargo aircraft fleet was the C-17 Globemaster III, which did not enter service until 1995; however, it possessed numerous improvements over its strategic predecessors. The USAF conceived it as a hybrid strategic-tactical airlifter. The aircraft operates primarily from a regular airfield, but the Air Force designed the airlifter to take off and land on shortened and austere airfields. It also had, unique among the cargo (C) class, thrust reverses that enabled it to back up without having to turn around by moving forward. Also different from its sister aircraft, the pilot or copilot flew it through use of a center stick, rather than the traditional yoke of heavy aircraft.

The easiest way to visualize a cargo aircraft's capacity is through the Air Force's standardized 463L pallet system. Each pallet represents a uniform area of cargo space, roughly nine feet by seven feet and capable of carrying up to 10,000 pounds each. The C-130 has six 463L pallet positions; the C-141, eighteen; the C-5, thirty-six; and the C-17, eighteen. In terms of sheer cargo capacity, the C-5 dwarfs all of the other "C" aircraft, even carrying twice as much as the newer C-17; however, the C-17's ability to provide tactical airlift freed up the C-5 fleet to focus exclusively on the strategic airlift function for which it was designed. Even the Air Force's KC refueling aircraft accommodate a certain number of pallet positions for the movement of cargo in addition to aviation fuel.

The fact that KC (air-to-air refuelers) also fall under the rubric of air mobility provides a nice segue for the discussion of the development of the Air Force's refueling aircraft: the KC-135 Stratotanker and the KC-10 Extender. Earlier in this work, I briefly mention the "Question Mark" flight in which airmen kept an aircraft aloft for more than 150 hours in 1929, but even the Air Force's own Air Mobility Command later admitted that flight "portended little militarily." The feat was indeed impressive in its day, but it was a long way from the ability to rapidly and safely sustain air-to-air refueling operations—something that did not become commonplace until the post–World War II era. In fact, it was the Cold War, the development of "jetomic" bombers, and the fear of nuclear holocaust that provided the impetus for the development of aerial refueling as a mission of the USAF. While the B-36 could fly from the United States to Europe without refueling, the follow-on B-47 and B-52 could not. The Air Force decided all of its future bombers needed to be capable of aerial refueling.[5]

Thus, between 1954 and 1964, the USAF procured more than 700 KC-135 Stratotanker aircraft. Nearly 400 of these remain in service between the active, reserve, and guard components. The follow-on to the KC-135, the KC-10 Extender, entered service in 1981; however, only 60 of these were ever procured with 59 remaining on the active inventory. As this book goes to press, the

USAF is currently fielding its next-generation refueler and transport, the KC-46 Pegasus. The aerial refueling fleet coupled with the cargo fleet represent the lynchpin of tactical airpower. All tactical and operational success flows from the Air Force's ability to move cargo and refuel its aircraft in flight.

Grenada and Panama

Three events merit brief mention in relation to the manner in which the Air Force returned to a useful way of war throughout the 1980s. Two of these, the invasions of Grenada and Panama (in 1983 and 1989, respectively), are of relatively little significance, in order of magnitude of Air Force operations, but the third, Operation El Dorado Canyon, represented a clear departure from conceptions of how to use airpower in the 1960s and 1970s. The bombing of Libya in 1986 clearly demonstrated that tactical assets now had the ability, the duration, and the payload to carry out decisive strategic attacks. Each will be explored in turn.

In the wake of the capture and assassination of Grenada's Marxist-Leninist prime minister Maurice Bishop and the ensuing coup d'état, the Regan administration, expressing concern for the hundreds of American medical students on the island, launched Operation Urgent Fury against the Caribbean island. Coming just four years after the Iran hostage crisis, the fear that a large number of American students might be captured made a compelling case for quick action. Although the country of Grenada had no air force, during the US invasion of the island in October 1983, the American Air Force believed that Cuban combat aircraft might attempt to disrupt the operation, and US military planners knew Cuban military personnel were on the ground on the island. Therefore, F-15s "patrolled the Caribbean Sea north and west of Grenada to detect and deter any air or sea movements from Cuba."[6]

At the same time that the F-15s patrolled between Cuba and Grenada, other Air Force assets overflew the island of Grenada to provide intelligence for senior leaders in the Pentagon as well as to tactical-level reconnaissance to troops on the ground. This included MC-130s and U-2s. Close to the ground and in direct support of US troops, the USAF provided CAS with AC-130s and A-10s. The F-15s played a defensive role while Air Force Special Forces aircraft and some accompanying A-10s provided close air support in support of the actual invasion.[7]

USAF EC-130s conducted command and control and psychological operations missions. At the same time, once the military forces secured their initial objectives, Air Force airlift assets—in this case, C-130s, C-141s, and a single C-5—conducted noncombatant evacuation operations.[8]

While not exactly a ringing victory against a worthy adversary, Operation Urgent Fury proved successful and accomplished the goals laid out by the

Reagan administration. The American medical students were rescued, and the burgeoning Communist threat to this small Caribbean country was eradicated. One historian noted, "Despite mistakes from which the defense department and the Air Force learned valuable lessons, URGENT FURY was unquestionably a success." However, and perhaps more important, the invasion was "the first clear U.S. military victory since the war in Southeast Asia, restoring pride in the United States and its armed forces that had declined in the wake of setbacks in Vietnam, Cambodia, Iran, and Lebanon." The USAF was making a comeback of sorts from the experience of Vietnam. The swagger of the American aces of yore was slowly returning to America's tactical air force.[9]

In December 1989, the United States launched Operation Just Cause (early planning was done under the name Operation Blue Spoon), in order to arrest Panamanian leader Manuel Noriega. President George H. W. Bush is reported to have given the operation the go-ahead by saying, "Okay, let's do it. The hell with it!" As in Grenada, USAF fighters provided support and protection from any Cuban interference. Fighters also protected the assaulting forces. Messages from headquarters Air Forces Atlantic (AFLANT) ordered four F-15 aircraft from the 33rd Tactical Fighter Wing placed on alert. Each F-15 was armed with four AIM-9s and four AIM-7s configured for use in a combat air patrol (CAP) to protect other airborne assets, but with the rather ambiguous orders: "Peacetime ROE [rules of engagement] applies." Other Air Force assets included EF-111s, EC-130s, and E-3s. Of particular note, the invasion saw the first combat use of the low observable F-117 Nighthawk.[10]

Airlift led the way, protected by the F-15 CAP, and moved two battalions of the 82nd Airborne Division on twenty C-141s. This was but a small portion of the overall airlift effort taken to get the attacking force into Panama. Two hundred aircraft took part in the operation. The Joint History Office of the JCS lists "80 C-141s, 22 to 25 C-130s, and 11 C-5s" along with an unspecified number of KC-135s and KC-10s, again demonstrating the importance of airlift to USAF, and joint, operations.[11]

Overall, senior leaders inside the Department of Defense heralded the mission as another success and a milestone in the ability of the US armed forces to work "jointly." A message from the commander of the 18th Airborne Corps, Lt. Gen. Carl Steiner, stated after cessation of hostilities: "This was a true Joint Operation all the way, just the way it's supposed to be done. We validated our joint training and procedures, and justified the nation's trust in our ability to rapidly employ Joint Forces in support of US interests."[12]

Operation Just Cause was not without its critics. The first combat use of F-117s came under particular condemnation after their bombs hit not an enemy barracks, but rather an adjacent field. An official history pinpoints the dichotomy in having the F-117 pilots attack a field: "Meanwhile, the pilots of the two F-117As flew to Rio Hato to drop one 2,000-pound bomb each within 150 yards of the PDF's [Panama Defense Forces'] 6th and 7th Rifle Company barracks to stun and confuse the occupants just before Rangers of Task Force RED parachuted into the area." A page later, it notes "the troops of the PDF's 6th and 7th Rifle Companies managed to overcome the shock of the F-117A strikes and fight for over five hours before 250 surrendered." Although this is completely conjectural, there remains the counterfactual possibility that the Panama Defense Forces would not have demonstrated such resolve in defense of their homeland had the bombs struck the barracks.[13]

Brig. Gen. Walter Worthington, at the time vice commander of the 12th Air Force and US Southern Air Forces, drafted a statement for his boss, Gen. Maxwell R. Thurman, the US Southern Command commander, rebutting the idea that the F-117s missed their targets. Worthington routed his statement through TAC headquarters for approval as well. It reads, in part: "As you know there has been a controversy about the results of the F-117A mission. To answer your specific question on the circular error of probability of the two bombs dropped by the F-117As, the answer is that the bombs hit precisely where they were aimed. This is based on detailed analysis of mission tapes. However, I must tell you that the aim point used by the pilots was slightly different than planned. The bottom line is that the F-117As dropped their bombs in the area near the barracks as directed to minimize actual destruction and loss of lives."[14] It was General Thurman who had advocated to Gen. Colin Powell and Secretary of Defense Dick Cheney for both using the F-117A and bombing near the barracks.

El Dorado Canyon

In between Grenada and Panama was another operation that proved a bit more conclusively how the Air Force's concept of warfare was morphing into preeminently tactical operations while still having strategic effects. Perhaps no operation better demonstrates just how much air warfare had changed in the preceding decade than the strikes code-named Operation El Dorado Canyon against Mu'ammar Gaddhafi and the Libyan military. The 48th Tactical Fighter Wing, stationed at Lakenheath Air Base in the United Kingdom, was to lead the Air Force's part of the attack against Libya. The wing had recently attended two training exercises and as well as Combat Hammer, a training exercise where pilots employ live ordnance against targets. Again, contrary to the practices during the Vietnam War, American fighter pilots trained as they would fight and entered into combat scenarios with a much better appreciation of their weapons systems and their own combat capabilities.

The mission planning and deception operations for El Dorado Canyon were extensive. The F-111Fs were loaded and armed inside hangers. The aircraft employed radio silence procedures except for a brief status check after launch. The base filed no flight plans. Takeoffs used a light gun instead of gaining clearance through the tower. From there, it was a simple matter of extensive air-to-air refueling and a grueling flight around the continent of Europe (to remain in international airspace), through the Strait of Gibraltar, and on to the targets in Libya.[15]

The threat along the Libyan coast was formidable. It certainly represented the most sophisticated IADS encountered by USAF aircrews since Vietnam. Libya's IADS included SA-2 Guidelines, SA-3 Goas, and SA-6 Gainful batteries as well as associated AAA—all built in overlapping and concentric circles intended to defeat an attack from the air.[16]

The US Navy and USAF worked closely to determine the right composition of the strike package. The Air Force tapped the 48th Wing's F-111F Aardvarks for the mission. Capable of carrying an assortment of weapons internally or externally and operating in all-weather environments with the ability to use its precision-strike weapons from low altitude, the Aardvarks were tailor-made for such long-duration, endurance missions. The F-111s attacked targets around Tripoli (Bab al-Azizia, Sidi Bilal naval training complex, and military aircraft parked at the airport) while Navy aircraft attacked targets in Benghazi (Jamahiriya military barracks) and also provided the CAP should the Libyan Air Force decide to engage with the attackers. The attack was successful and resulted in only one friendly loss (Karma 52).[17]

El Dorado Canyon proved several things about air warfare in the 1980s, although it seems clear these were lessons learned directly during the Vietnam conflict. First, tactical assets, in this case the F-111Fs, were capable of producing strategic effects, once believed to be the domain of heavy bombers. Second, the event combined fighter-bombers and aerial refuelers over a scale and distance never seen before and demonstrated the importance of aerial refueling operations.

The Interwar Period as Learning Environment: Aggressors, Red Flag, and the Weapons School

Perhaps the biggest change in how the Air Force prepared for and conducted war was seen in the areas of training and employment of fighter aircraft. This began with the creation of an exercise dedicated to simulating combat and ended with a conflict that validated the exercise as worth the loss of lives that occurred in training. In the 1970s and 1980s, the USAF fixed the problems experienced in Vietnam by providing its pilots—starting with fighters, but expanding to all combat fliers—a verdant environment and a paradigmatic shift in training that greatly improved Air Force preparedness for combat.

This Fundamental change to the Air Force's way of war, especially to how it trained for war, began to occur in 1972. In that year, the USAF established the first of the Aggressor squadrons. These squadrons were equipped with T-38s and later F-5 aircraft that more closely resembled the shape, size, and performance of enemy aircraft—akin to the MiG-21 employed by the Soviets and Soviet bloc countries. The purpose of the Aggressors was to employ their fighters in a manner similar to Soviet pilots and thus to expose US fighter pilots to aircraft that were smaller and more nimble than their own and to demonstrate as closely as possible Soviet tactics in air combat. Aggressor units traveled the country to Air Force bases to effect to teach American fighter pilots Soviet tactics.

Further changes occurred in 1975 with the establishment of the Red Flag exercise. Rather than having the Aggressors travel to where the home station pilots might have a "home field advantage," TAC created an exercise where pilots "deployed" to Nellis Air Force Base in Nevada for a two-week-long air combat exercise. Detailed in my book *The Air Force Way of War*, Red Flag improved combat capability and continues to expose aircrews to "realistic training" nearly a half century since it began.

All the while, changes were also taking place at the Fighter Weapons School. Gone were the days of "Forty-Second Boyd," wherein the infamous Col. John Boyd bragged about his ability to transition from a defensive position to one of offense against any pilot in under forty seconds. Replacing this was the motto "Humble, approachable, credible." The Weapons School took relatively junior officers and trained them to be the most proficient operators of their particular airframe. Graduates received a patch that boldly denoted them as a "Graduate: USAF Fighter Weapons School," worn on the left side of their flight suits. A graduate of the Weapons School became a "tactical systems expert" in his or her particular airframe.[18]

The Air Force Weapons School, initially activated in 1965, traced its roots back to the Air Force Gunnery School of the 1940s and 1950s. In May 1950, the USAF released Air Force Regulation 53-6 to teach "fighter gunnery, rocketry, and dive-bombing at a high standard of proficiency." The gunnery school had a single entry requirement: "Be currently proficient in the piloting of jet fighter aircraft."[19]

The idea of a unit dedicated to the employment of an aircraft as a weapons system began in 1952 when the words "Combat Crew" were dropped from the name of the school and replaced with "Fighter." By 1953, the instruction of fighter pilots in the art of weapons employment was underway at Nellis. By 1954, it was known Air Force–wide as the Fighter Weapons School. As aircraft entered the Air Force, instruction on their use was added to the curriculum at the school. For example, in 1963 Nellis Air Force Base added a special course for the new F-100D/F models. As the war in Vietnam heated up, the Weapons School responded by standing up a new division devoted to "the improvement of US tactical air capabilities in Southeast Asia and the future." Such training continued to expand as Nellis added new missions including the Aggressor squadrons and the Red Flag exercise, all of which fell under the umbrella of the USAF Fighter Weapons Center.[20]

Since this change in approach, the Weapons School has undergone a massive expansion and moved beyond the training of fighter and bomber pilots to include cargo, refueler, space, and cyberspace weapons officers. By 2020, the school contained nineteen different squadrons operating at Nellis and nine other locations around the country. The school graduated around one hundred graduates in each of its biannual classes. The graduates of the USAF Weapons School then went back to a flying squadron, or equivalent unit, where they served as weapons officers.[21]

These three events—the change in approach at the Air Force Weapons School, the creation of the Aggressor squadrons, and the beginning of the Red Flag exercise—made Nellis Air Force Base the "home of the fighter pilot" and one of the most desirable locations for a fighter pilot to be stationed (that the city of Las Vegas was just down the road certainly did not hurt).

By the mid-1980s, pilots considered these realistic training exercises the norm and something that all fliers participated in as part of normal stateside operations. The post-Vietnam changes encouraged a renaissance in the study of fighter combat. For USAF pilots, such study began not on the Nellis training ranges, but much earlier in their career. For those tracked into fighters, it began with the T-38 Talon. Here a new pilot learned the art and science of basic fighter maneuvers (BFM). BFM involves both offensive tactics (being in a position of advantage over the enemy aircraft) and defensive tactics (being in a position of disadvantage to the enemy). More than that, though, were courses in formation flying and instruments and navigation. It was, and remains, a highly technical career field. While the historical mythology of the fighter pilot demonstrated cunning and luck, training in BFM emphasizes geometry and physics. Terms like "angle of attack," "radius/rate of turn," and "relative range" were the order of the day.

This renewed emphasis on the importance of tactical airpower manifested itself in other ways as well. Pilots learned both the art and science that was fighter combat. Key to this art, pilots learned to place their aircraft in the "offensive" position. Being in an offensive position was defined as being behind the enemy aircraft, which pilots called the enemy aircraft's—or "bandit's"—3-9

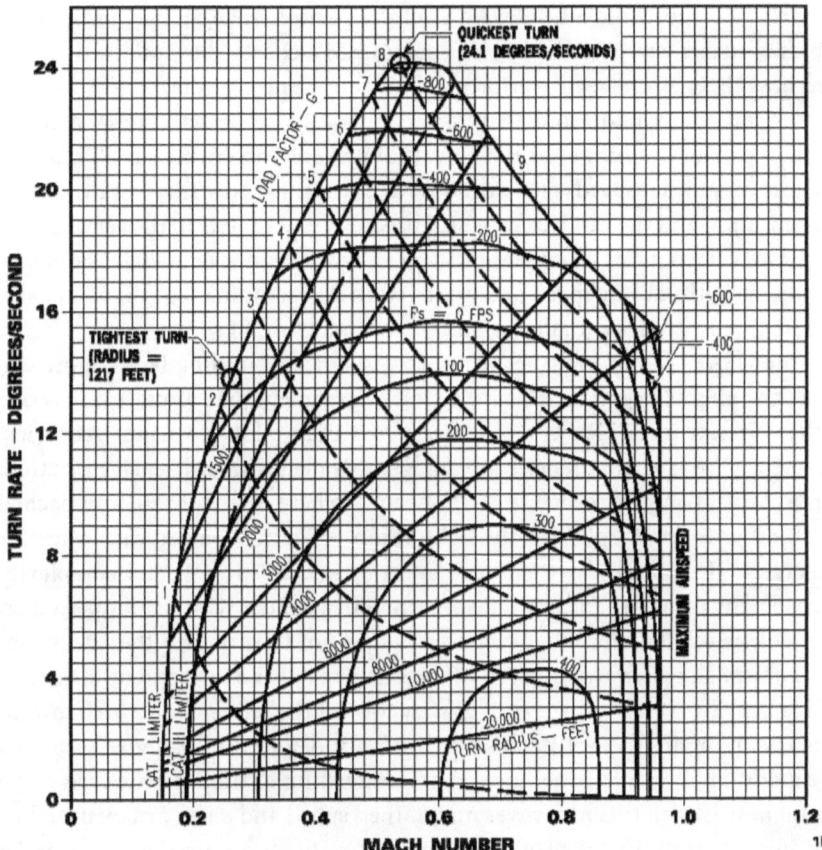

Energy maneuverability (EM) charts demonstrate aircraft performance at a given speed (Mach number), altitude, and turn rate resulting in a calculation of quickest turn, tightest turn, and so on, as shown in this F-16 EM chart. An enemy aircraft's EM chart can be overlaid to determine maximum advantage in any engagement. EM chart image courtesy of the US Department of Defense.

line. The 3-9 line is the imaginary space located behind the bandit's wings (like the 3 and the 9 positions on a clock). Pilots achieved getting behind this line either by the traditional stealth and cunning seen during World War I, through luck if the enemy overshot and flew in front of you, or through the skill of maneuvering. Whatever way, fighter pilots had one mission: kill the bandit.

Pilot training and tactics developed in other ways. John Boyd's energy maneuverability theory (EMT), developed in the 1960s, had a profound impact on air-to-air combat. The EMT charts Boyd helped develop were unique for each aircraft and detailed what a particular aircraft's performance would be at a particular speed and altitude. An aircraft's performance changes as the variables of speed, energy, and altitude change. EMT charts allow pilots to visualize how their aircraft will perform against an enemy's aircraft and where an American fighter pilot would have an advantage in the arena of aerial combat. A pilot can look at his or her EMT chart and overlay an enemy's EMT chart to determine at what speeds and altitudes were advantageous. This was invaluable information and the EMT chart is still used by fighter pilots in the aerial arena. Boyd described it this way: "It allows you to define maneuverability. The ability to change altitude, air speed and direction in any combination." Boyd's charts aided greatly in conceptualizing and then conducting BFM.[22]

The USAF trained pilots for aerial combat with a building-block approach. Mock dogfights began from the simplest engagement: one aircraft versus one other aircraft (1v1). Flight instructors tested each pilot in various scenarios: where a pilot had an offensive advantage, was placed in a defensive position, or a neutral engagement (head-on-head) where the fighters approached each other in level flight and the fight began when they passed each other, or "merged." This level was then built upon by adding aircraft and complexity to each engagement. For example, in a 1v1 engagement with a pilot placed on the defensive, he (later s/he) might be taught not to turn into the attacker at maximum G, which would bleed off precious speed and energy, but instead to place his "lift vector" or center of gravity toward the "enemy" and turn toward the enemy in an attempt to close the distance before the "bad guy" was in a weapons employment zone. Instead of committing to the smallest turn circle or highest-g turn, this maneuver maintained speed and energy maneuverability after the turn. In a neutral engagement, two fighters passed head-on and a pilot faced the choice of either entering into a turning fight with his bandit or "blowing through." If the trainee decided to turn, and depending on the turn of the "enemy" fighter commonly known as "red air," either a "one circle" or "two circle" engagement began. From there, complexity increased. As tactics changed, a method was needed to get the word out. Fighter pilots would never accuse themselves of being overly intellectual, but they certainly did their fair share of reading to keep up with the changes occurring in their community.

Exercises at home stations and other locations, the most famous being Red Flag, encouraged pilots to expand their knowledge beyond BFM into air combat maneuvering (ACM) and air combat tactics (ACT). The Fighter Weapons School at Nellis Air Force Base became the finishing school for a master's degree in tactical employment of a particular airframe. One of the methods for passing on the changes in tactics and doctrine from Nellis to fighter pilots around the world was the *Tactical Analysis Bulletin*.

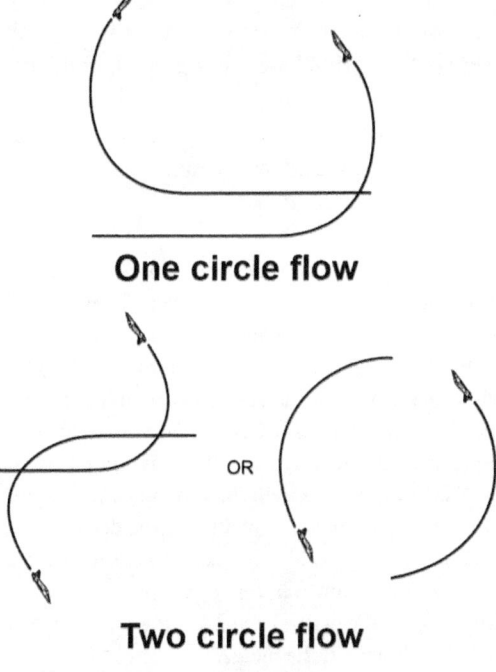

One circle flow

Two circle flow

This flow turn chart demonstrates how, in a standard 1v1 head-on-head engagement, aircraft will turn toward each other, creating a "one circle" fight, or away from each other into a "two circle" fight. Flow turn chart image courtesy of the US Department of Defense.

The *Tactical Analysis Bulletin* covered a gamut of tactics (friendly and enemy), techniques, and procedures. It was equivalent to a peer-reviewed journal and was written at the classified level; many of the authors held a "PhD in airpower" by virtue of their having graduated from Fighter Weapons School and their teaching on the faculty there, hence the articles ranged from broad esoteric discussions on airpower employment at the tactical level to highly nuanced articles on employment of a particular weapons system by the expert in the field. For example, the April 1986 *Tactical Analysis Bulletin* includes an article focused on "acknowledging and defining the increased capabilities and advances in Soviet technology," and advocating knowledge of same as "the first step in improving our own training programs."[23]

The article went on to state that "the Soviets or any potential threat may be able to configure their older jets with advanced 'point and shoot' and 'wish you were dead' missiles." It also detailed the "golden rules" of BFM, including "Lose sight, lose the fight" and "Maneuver in relation to the bandit." It also stressed the importance of "Nose position vs. energy," stating: "Energy is used to gain nose position. Nose position is used to gain lethal weapons' parameters." In summation: "BFM is used by the pilot to place himself in a piece of sky from which he can launch lethal ordnance, or to keep from becoming a star on the side of somebody else's jet."²⁴ It was an exciting time for new fighter pilots in the United States Air Force. They were better trained, better equipped, and better prepared than any that came before them, but it was still a deadly business.

The essence of air-to-air combat had not changed since the days of Eddie Rickenbacker even though the altitudes, speed, and performance had advanced dramatically. A *TAB* article read by fighter pilots stated:

> In the air-to-air combat arena, engagements can be described as fighters established in the same turning circle and fighters outside each other's circle. Turning circles are a function of turn radius and turn rate, which are functions of airspeed, G, and power. The classic BFM engagement puts one fighter trying to get inside the other fighter's turn circle (offensive) or one fighter trying to keep the other fighter out of his turn circle (defensive). Of course, a third case exists where both fighters are trying to get in each other's turn circle at the same time (neutral). The concept of turn circles is critical in all three BFM situations. It is best understood as the offender by analyzing your position relative to the defender's present turn rate. If the defender's present turn rate will place you ahead of his 3-9 line, then you are outside his turn circle. However, if you can make the turn with him, you are inside his turn circle. In the neutral situation, the turn circles are defined by the initial moves and lead turns at the merge.²⁵

In an article titled "To B(FM) or not to B(FM): That Is the Question," Lt. Col. Lynn O. High detailed the changes that were made in the 1970s: "In 1972, when the Aggressors were formed and DACT [dissimilar air combat training] became a word you could say openly the biggest deficiency we had in air-to-air capability was human performance. Our tactical BFM skills were weak.... By 1975 DACT and four Aggressor squadrons equipped with new F-5Es were the hottest game in town. And what did those Aggressors do? They taught BFM.... Even as Soviet tactics simulation quickly grew as an Aggressor mission, BFM was still the heart of every debriefing."²⁶

Air Combat Maneuvering

Learning basic fighter maneuvers was only an introduction. Sure, fighter pilots considered it the best part of their job. It was at the heart of fighter pilot mythology, the purest form of the "knights of the air" mentality. Being a fighter pilot engaged in BFM pitted man versus man (and after 1992, women too) in a fight to the death (or at least a simulated kill). But all that was what you would expect—*basic* fighter maneuvers. After mastery of BFM, maneuvers became more complicated, specifically ACT and ACM. The ever-growing challenges taught pilots to think dimensionally and spatially about the environment around them, as 1v1 exercises morphed into 2v2 and 2vX (an unknown number of "enemy" aircraft). Throughout the 1980s, fighter and bomber pilots used the methods and knowledge emerging from Nellis to ensure that American fighter pilots the world over were the best trained and most informed. The tactical and technical aspects discussed above represented how the human element interacted with and developed together with the technological.

All of these events, schools, and exercises were used to train fighter pilots to clear the skies over the battlefield and give bombers or air-to-ground attackers a more permissive environment in which to operate. However, one advancement in particular, occurring at the same time, demonstrated that sometimes a single technological advance could achieve a different method entirely of "over not through."

Low Observability

The advent of low observable technology, more commonly referred to as stealth, came in the mid-1970s through a cooperative effort between the Defense Advanced Research Projects Agency (DARPA), Lockheed Martin, and the US Air Force. The concept was simple enough. An aircraft with the right design features and coatings could significantly reduce its own radar return against an enemy's radar, leading to an aircraft that could pass through enemy territory unobserved. It is important to remember that "stealth" never meant invisible; it just meant that an aircraft with the right low observable characteristics became prohibitively difficult for enemy radars to find, fix on, and engage. Thus, a low observable aircraft could enter enemy territory and depart faster than enemy radar could pass along the necessary information to destroy it or intercept it.

The initial DARPA project went by the code name Have Blue and resulted in the production of two low observable aircraft: *Have Blue* HB1001 and HB1002. It was also derisively called the "hopeless diamond," given its shape and nonaerodynamic attributes. Although both aircraft were lost in crashes (both pilots survived the incidents), the Air Force was pleased with the results and contracted Lockheed Martin to produce the aircraft under the new code-word-protected program: Senior Trend. The Have Blue aircraft design evolved into the F-117 Nighthawk.

The F-117 was larger than the Have Blue aircraft, but it retained much of the multifaceted shaping of its predecessor. Notable differences included not only the size, but also the vertical stabilizers that, when viewed from the front or rear of the aircraft, made a very observable "V" shape in the F-117, as opposed to the Have Blue model's stabilizers, which cantered toward each other. The two primary design considerations in the F-117 were the shaping of the aircraft and its coatings. The first used a computer program called "Echo" to design the panels that forced radar waves to "bounce" up and away from the aircraft rather than send a strong return signal. The second consideration was the radar-absorbing material (RAM), which, as the name implies, absorbs radar waves rather than reflecting them.

The Air Force operated the F-117 out of Tonopah Test Range, the same airfield that also housed the Air Force's secret MiG squadron. Tonopah, located within the Nevada Test and Training Range, was conveniently a short flight from both Nellis Air Force Base and Area 51. The pilots who flew the F-117 were "stationed" at Nellis and purportedly flew the A-7 Corsair. Each week they flew out to Tonopah and flew the F-117 during hours of darkness. The F-117 entered combat during Operation Just Cause, but truly entered the American lexicon as the stealth fighter during Operation Desert Storm.

Conclusion

The "interwar period" between the Vietnam War and Operation Desert Storm was a golden age for the Air Force. Increased budgets, personnel, and training opportunities, coupled with the procurement and fielding of the next generation of aircraft, better positioned the organization for the combat it was soon to face. The trials and errors made during Vietnam had given way to a new generation of officers trained by those who experienced the hard-knock lessons firsthand. Although Operation Eagle Claw, the attempted rescue of Americans held during the Iran hostage crisis, proved disastrous, the special operations community would learn lessons that influenced their training and actions for a generation. For the line Air Force, Urgent Fury, El Dorado Canyon, and Just Cause had a profound influence. All three of these operations provided examples that the Air Force was capable of operating in joint environments, and the Libya raid proved that tactical fighter-bombers were capable of similar strategic effects as heavy bombers. Finally, exercises including Red Flag, developments in programs like the Weapons School, the creation of Aggressor squadrons, and home station continuation training all produced pilots better equipped and better trained than at any other time in the history of the Air Force. While this primarily applied to the tactical community, advances had also been made in the bomber and airlift (both cargo and refueling) communities. It was, to use a clichéd phrase, a perfect storm, that came to pass at a most fortuitous time.

CHAPTER 7

DESERT STORM AND THE BALKANS CAMPAIGNS

You have waged the most successful air campaign in history.... We expected and accomplished more than many thought possible.

—*Brig. Gen. Buster Glosson to Desert Storm aircrews*

At 0238, local time, 17 January 1991, US Army AH-64 Apache helicopters—guided to their targets by US Air Force MH-53 Pave Lows—unleashed a barrage of missile, rocket, and cannon fire into a pair of Iraqi early warning radar sites and effectively created a "black hole" in the Iraqi military's ability to detect incoming aircraft. Twenty-two minutes after the destruction of the early warning sites, low observable F-117 Nighthawks delivered laser-guided GBU-27 bombs on targets in downtown Baghdad. Into the black hole created by the Apaches flew all types of fighters, bombers, and fighter-bomber aircraft belonging to the USAF, the US Navy, and the United States Marine Corps. Leading the way were twenty USAF F-15C Eagle air-superiority fighters. Flying from bases in Saudi Arabia in four-plane formations, these fighters turned left into western Iraq flying line abreast and creating a moving wall of airpower, leading the way to intercept any enemy aircraft that might try to hinder the attacking F-15E Strike Eagles following in their wake to strike targets in western Iraq.

Elsewhere over Iraq, fighters and attack aircraft belonging to the US Navy—including A-6 Intruders, A-7 Corsairs, F-14 Tomcats, and F-18 Hornets—executed SEAD operations and CAP missions. Joining the American aircraft were others from the air forces of the United Kingdom, Canada, France, Saudi Arabia, and other coalition members. From an air base in Guam, some 10,000 miles away, gigantic B-52 Stratofortresses dropped cratering munitions on airfields. Elsewhere in Iraq, EF-111 Raven electronic jammers forced Iraqi radar operators to turn their radars to full power settings in order to see through the electronic haze forced on them by the "Spark Varks." This allowed USAF F-4G Phantom II "Wild Weasels" to shoot AGM-88 HARM (high-speed antiradiation missiles) directly into the radars.[1]

Amid the well-choreographed dance of attacking aircraft, confusion reared its head. Aircraft calling out strikes, maneuvers, vectors, and other information jammed the radios, causing "sensory overload" to many of the pilots. One of the F-15Cs piloted by Capt. Jon Kelk (Pennzoil 63) fired an AIM-7 Sparrow missile and killed an Iraqi MiG-29, but he did not receive confirmation of his aerial victory for several hours. Capt. Rob Graeter (Citgo 51) did not have to wait for his victory confirmation. He fired an AIM-7, which tracked his MiG and exploded; Graeter witnessed the MiG enter the missile's shrapnel and "it just disintegrated. There were burning chunks of aircraft going in every direction."[2]

Throughout the sky over Iraq, a lopsided air-to-air melee emerged. USAF Capt. Steve Tate killed a Mirage F-1; the Navy's first aerial victory in Desert Storm came from Lt. Cdr. Mark Fox flying from the deck of the CVN-60 USS *Saratoga*, a veteran carrier that saw service at Yankee Station during the Vietnam War. Lieutenant Commander Fox fired a pair of missiles: an AIM-9 Sidewinder followed by an AIM-7. As both missiles found their MiG-21 target, Fox watched as the remains of the burning MiG passed a thousand feet below him. As Fox watched his aerial victory pass beneath his cockpit transfixed by the sight, he was jarred back to the present when he heard through his headset "Splash One!" as another member of Fox's flight, Lt. Nicholas "Mongo" Mongillo, shot down another MiG-21, but it was Fox who drew first blood for the US Navy.

On 17 January 1991, aircraft of the USAF and USN destroyed nine enemy aircraft, six more were added on two days later, and over the course of the next two and half months, twenty-nine more enemy fighters and helicopters would be destroyed in air-to-air combat. By clearing the skies of enemy fighters, the first wave of fighters allowed strike aircraft to systematically destroy Iraq's air defense system. Surface-to-air missile sites, radar sites, airfields, antiaircraft artillery sites—all part of a dense integrated air defense system (IADS)—were taken apart piece by piece, giving the multinational coalition air supremacy. This was all part of the aerial plan for Operation Desert Storm: gain and maintain air superiority by ridding the sky of the Iraqi Air Force and destroying the multilayered IADS to give coalition fliers freedom of movement through the air.[3]

For the most part, the fliers in Iraq that night were able to keep their hands on their flight control sticks and their throttles, both of which were equipped with dozens of buttons and switches that allowed them to sort, select, and fire weapons all without having to remove their hands from these devices. This arrangement, known as "hands on throttle and stick" (HOTAS), was a direct fix to a situational problem encountered by fliers in Vietnam: pilots and aviators, in the heat of aerial combat, had to reach forward in their cockpit to flip missile select switches. Pilots operating the HOTAS never had to remove their eyes from their head-up displays (HUDs).

The weapons used during Desert Storm performed better than similar models used during the war in Vietnam. Early AIM-9 variants— the Navy and Air Force's "heat-seeking" missile—notably the "B" models used in Vietnam, were designed to be fired from the six o'clock position behind a nonmaneuvering target, envisioned as an incoming Soviet bomber, and not against small, nimble, high-g maneuvering fighter aircraft. The seeker head of these early AIM-9Bs failed to track targets moving at higher than 3 g. Other versions updated and improved the "B" variant deficiencies, including the AIM-9D/E/J models, but despite improvements, the AIM-9s used during the Vietnam conflict never reached an "all-aspect" employment envelope. By the time of Desert Storm, the US air fleets were equipped with the AIM-9M, which had a "greater tracking rate and tighter turn capability than earlier series missiles." This improved missile had a higher probability of kill (Pk) rate; in other words, it had a higher chance of hitting what it was fired at, plus its improved ability to discriminate a target against ground clutter and other heat sources meant it had a higher ability to lock on and stay locked on to its intended target. One pilot recalled that locking on to a target with an AIM-9 caused an audible growling sound to be emitted into the pilot's headset, and firing an AIM-9 was a matter of "Growl, shoot. No growl, don't shoot. Even a fighter pilot could understand that." Improvements made in the AIM-9L version that carried over into the AIM-9M meant that the missile could be fired at a target from any point on 360-degree circle or from "all aspects," including head-on. The same was true for the AIM-7 radar homing missile, maligned during the Vietnam War but successful in downing nearly twenty aircraft over Iraq. Much like the AIM-9 model updates, the AIM-7 Sparrows had also undergone significant technological evolution, just like the Sidewinders. Although every firing of a missile must be taken in context of varying factors—turn rate, speed, altitude—by the time of Desert Storm, the improvement efficiencies made in the twenty years since Vietnam on the AIM-9 and AIM-7 had increased their Pk rates to as high as 50 percent, and firing two missiles improved the probability of a kill to 70 percent. Pk rates of individual missiles are a relatively tactical approach to the efficacy of airpower in any given conflict; other factors played into airpower's successes and failures.[4]

Had the ghosts of the air wars in Vietnam been vanquished? Did Operation Desert Storm indicate that airpower had finally reached full maturity? Of course, Vietnam and Desert Storm were not perfect examples of "good vs. bad" uses of airpower. Other quantifiable elements that hindered airmen in Southeast Asia turned into advantages over Iraq—namely, the environment, weather, and geography. Air Force historian Richard Davis noted, "Unlike Vietnam where the rain forest proved a formidable opponent, the desert in Iraq and Kuwait would maximize USAF combat strengths and Iraqi vulnerabilities."[5]

As we have shown, the fighter and bomber pilots of Desert Storm were also better trained than their Vietnam predecessors. During the years leading up to Vietnam, aerial combat training—in both air-to-air and air-to-ground environments—was not a priority. DACT aerial training against an "enemy" that was different from your own aircraft was virtually nonexistent. The Navy did not have either a gunnery school or a weapons school (until 1969 for the latter). During the Vietnam War, the first time a pilot might take off with a full combat load under his wings was likely to be his first combat mission. Conversely, in the years following Vietnam, this fundamentally changed. Young captains in their twenties and the older majors in their thirties flying in Desert Storm had each flown numerous realistic training engagements at their home bases, at the Red Flag exercise at Nellis Air Force Base, and at other training events and exercises. Many had fired or dropped live weapons at Combat Archer and Combat Hammer.

Not only that—these pilots then came home to a hero's welcome. Some went on to further success in the skies over the Balkans in the late 1990s. Their *air* leader was a singular individual: Lt. Gen. Charles "Chuck" Horner commanded all US air assets flying missions during Operation Desert Storm, all of them: Air Force, Navy, Marine Corps, and coalition aircraft. Serving as the joint force air component commander (JFACC), Horner reported directly to Gen. Norman Schwarzkopf and was responsible for planning and executing the largest air campaign since the Vietnam War. In this position, which did not exist during that conflict, Horner was the "single manager" for all combat aircraft.

The air wars over Vietnam in effect helped foster the success of Operation Desert Storm. The air wars in North Vietnam, South Vietnam, Laos, and Cambodia were, in many respects, the opposite of what occurred during Desert Storm. The deficiencies of the former conflict were not lost on the planners of Desert Storm; they studied them and conceptualized their own war as the antithesis of Vietnam—in fact, the original name for the Desert Storm air campaign was Instant Thunder, a direct repudiation of Vietnam's Operation Rolling Thunder. Many of the problems of command and control that we will examine in this book were absent in Desert Storm, including "all the familiar

Vietnam elements—disorganization, terrible planning and insight, and inability to achieve a goal." Perhaps no one understood these issues better than Horner, himself a veteran of the Vietnam conflict. The experience of the loss of the war in Vietnam altered Chuck Horner and an entire generation of American airmen. After Desert Storm, historians and those who participated in the conflict expressed overtly how fundamentally different the two wars were from each other.

After Desert Storm, historian Larry Cable claimed, "The soul-sapping specter of Vietnam had finally been laid to rest."[6] Richard Hallion, a former head of the Air Force's History Office, stated that the United States entered Desert Storm with something it did not have in Vietnam: "Technology, equipment, logistics, doctrine, and training are all-important and necessary aspects of any military organization, but there are others that are at least equally significant: *experience and confidence*."[7] One of the key air planners for Desert Storm, Brig. Gen. Buster Glosson, remembered, "In Vietnam, nobody cared or bothered to try to fit it all together for us. . . . I was responsible for eight F-4 fighters . . . not a position to have a very broad overview of the war, but broad enough for me to realize that we never had a clue, at the unit level what our overall effort was trying to accomplish." Perhaps no one said it more succinctly than the man who came up with the initial air plan to be used against Iraq, Col. John Warden, when he told his Air Staff planning team, known as "Checkmate," in August 1990, "One of the things we want to emphasize right from the beginning is that this is not Vietnam! This is doing it right! This is air power!"[8] The success of Desert Storm could not have been achieved without the defeat suffered in Vietnam.[9]

The United States Air Force entered the final decade of the twentieth century well prepared to meet possible and perceived threats. An Air Force designed for combat in Western Europe saw the threat of Soviet invasion diminish and then recede entirely, but the great changes wrought in training and technology did not go unused. Nevertheless, it is doubtful had you asked anyone outside of the United States Central Command (CENTCOM) in 1990 where the next conflict was going to occur you would have received the answer of Iraq. And it is even more doubtful one would have foreseen that the Saddam Hussein's invasion of Kuwait in 1990 would set off a series of events that kept the United States locked in the region for the next three decades. In between Operation Desert Storm and Operation Iraqi Freedom would come several smaller operations in Iraq, but it was the larger ones that took place in the region known simply as "the Balkans" where it seemed as if airpower finally met all of the war-making expectations its advocates had initially dreamed it could. However, hindsight has proved a harsher judge of the age of airpower.

Williamson Murray stated in his book *The Iraq War: A Military History* that "the History of air campaigns is perhaps the most difficult for the military historian to depict, and the Iraq War is no exception."[10] It is certainly more complicated than Billy Mitchell's quip, "Air power [is] the ability to do something in the air."[11] Whereas it has long been a tradition in the US Army to conduct "staff rides," the same is not true of the US Air Force. Historians of airpower and aviation history struggle to re-create, for a given conflict, what happened in the air and what effects it had on the associated ground war. Still, the draw of being present where something happened—in this case, two somethings—is important to the aviation or airpower historian as well. In a not so strange twist of fate, some pilots got to "be there" ("there" being the skies over Iraq) twice.

Operation Desert Shield was a massive logistic undertaking. Operation Desert Storm saw the full use of the Air Force's combat potential. All fighter, bomber, and attack aircraft conducted operations on a massive scale, and not necessarily in the roles in which the aircraft had been envisioned. The tactical dominance of the Air Force was on full display as the USAF again demonstrated its ability to do many missions well: air superiority, strategic bombardment—this time with fighter and attack aircraft—and close air support were all conducted nearly simultaneously. After Desert Storm, the USAF entered a decade where this prowess was tested in the airspace over the Balkan Peninsula.

When Saddam Hussein moved his forces into Kuwait on 2 August 1990, the response from the international community was swift. Iraq was quickly condemned in the United Nations with Resolution 660 as even countries normally aligned with Iraq rapidly condemned the invasion as a flagrant violation of international norms. The United States military initiated actions in response almost immediately, in an effort as much to shore up Saudi Arabia in the event of a possible further Iraqi invasion as to deal with the invasion of Kuwait. While the Navy was moving two carrier groups into the Persian Gulf, the USAF dispatched forty-eight F-15s from the 1st Tactical Fighter Wing, out of Langley Air Force Base, Virginia, which landed in Saudi Arabia and began combat air patrols. The initial F-15s were followed by more F-15s from the 36th Tactical Fighter Wing in Bitburg, Germany. These were but the first of what eventually amounted to hundreds of combat aircraft, but none of these aircraft could operate in a vacuum. They needed pilots, aircraft mechanics, and ammo troops to arm them. These airmen needed food, sleeping accommodations, and further logistic support. To get all of this into theater, the USAF again used airlift operations to set the stage for the storm.

Airlift

Not enough has been written about the logistics of Operation Desert Shield. As William Head stated in his book *The Eagle in the Desert: Looking Back on*

U.S. Involvement in the Persian Gulf War, "Sadly, few experts, let alone people, are familiar with anything more than general aspects of what airlift, sealift, or logistics are all about. The same ignorance exists with regard to the role of these vital functions in the Gulf War."[12]

Airlift operations began within days of the invasion. On 8 August, a C-141 arrived carrying the lead elements of the US Air Forces Central (CENTAF) planning staff. It was perhaps inevitable, and yet probably unfair, that comparisons between Desert Shield and the Berlin Airlift would be made. For starters, the Berlin Airlift was an exclusively air operation as opposed to Desert Shield, in which sealift played a tremendous role. Second, comparing the C-47 to the C-141 or C-5 would be exactly like comparing a World War II fighter to a fifth-generation fighter; the two do not draw easy comparisons. However, it was not long before Gen. Merrill McPeak, Air Force chief of staff during Operations Desert Shield and Desert Storm, said it was "the equivalent of a Berlin Airlift every six weeks." This was stretching the truth a considerable amount, considering that when comparing overall numbers, Desert Shield moved 548,000 tons of cargo in nine months, while the Berlin Airlift moved 1.78 million tons in fifteen months. If anything was worth comparing, it was actually a smaller number—namely, that it took only 19,700 missions to move the cargo for Desert Shield as compared to the 189,960 missions during the Berlin Airlift. Still, it is not unfair to say that "Desert Shield soon became the most massive airlift in the history of air power."[13]

What is even more amazing is that the bulk of men and material was moved in such a short period of time. Two periods of time might be more accurate. The principal defensive network moved in August and September, with a second move of additional troops composing the offensive arm of the military response moved into place in November. To be fair, nearly 95 percent of the resources, equipment, and supplies were brought in by sealift, but 99 percent of the personnel moved through the air.[14]

What type of airpower did the USAF bring into the region? The simple answer is a smattering of everything: from Vietnam-era aircraft to the most recent additions to the force. These aircraft were capable of filling a variety of roles and missions, and the Air Force found missions for all of them. Some examples include the following: for the air-superiority role, F-15Cs; for strategic attack, the F-117 Nighthawk; for other air-to ground roles, such as CAS, interdiction, SEAD, and attack of forces in the field, F-15E Strike Eagles, F-111F Aardvarks, A-10 Warthogs, F-16C Vipers, F-4F Phantom IIs, and the B-52G BUFFs. Complimenting this offensive arm were dozens more air-to-air refueling aircraft (KC-10s and KC-135s) and search and rescue rotary-wing assets (MH-60 and MH-47). Still other aircraft included cargo, distinguished-visitor transportation, and intelligence, surveillance, and reconnaissance (ISR) assets. The list went on and on; and this is to say nothing of the aircraft owned and operated by both the US Navy and US Marine Corps or other rotary-wing aircraft operated by the US Army. A force fielded in the long wake of Vietnam was now preparing for a war against Iraq.[15]

In total, over seven hundred aircraft of all types and mission sets deployed around the Arabian Peninsula, primarily in Saudi Arabia, but also in the United Arab Emirates, Qatar, and Oman. Refueling and bomber aircraft were stationed farther away, including in Turkey and Spain and on Diego Garcia.

Planning the Air War: Instant Thunder

The planning for the air war began in earnest when Gen. John M. Loh answered the phone in his Pentagon office. On the other end was General Schwarzkopf, looking for the Air Force chief of staff, Gen. Mike Dugan. Dugan was not in the office, so Schwarzkopf informed Loh that he needed some help. From Schwarzkopf's point of view, his CENTAF planners, including Lt. Gen. Chuck Horner, were too busy overseeing the logistic ingress of troops and aircraft in order to deter further Iraqi aggression to truly plan for an air campaign if one were going to be needed in the future. It has not been said often enough that Schwarzkopf's phone call to the air staff was something of a foul. First, it cut his own air component lead out of the planning cycle; and second, the Goldwater-Nichols Act had been instituted in part to prevent this very occurrence of Pentagon planners interfering in combatant commanders' planning authorities. Still, it proved a fortuitous moment for the legacy and history of the USAF, and the senior air planner, Col. John Warden, would later call it that "wonderful call from Schwarzkopf."[16]

Despite this, the request for assistance eventually landed on the desk of Col. John Warden, who headed the Air Force's "Checkmate" division, a group of planners whose mission was to "red team" an enemy's actions. It might be said the genesis of the Desert Storm air war began in a stateroom on a cruise ship. When Iraq invaded Kuwait, Warden was on vacation with his family. He immediately started thinking about how he should respond if asked for his opinion. Having recently reviewed the CENTCOM operations plan 1002, he was already familiar with the plan and response as it existed on paper, but he certainly had opinions about how to improve it. By 8 August, Warden was back in his Pentagon office with roughly three dozen air planners, and they set to work devising a plan for defeating the Iraqi Air Force and ejecting Saddam and his forces from Kuwait. Warden's ideas represented the very nature of "over not through" that would culminate in a victory through airpower.[17]

Warden was something of an intellectual force inside the air force. He was, like most of the senior air leaders at the time, a veteran of Vietnam and had focused his intellectual energy on adapting the lessons learned since the end of that conflict. During his year at the National Defense University, he wrote a thesis, published two years later in book form titled *The Air Campaign*. It would become one of the principal works on airpower theory. Warden described his work this way: "*The Air Campaign* is an attempt to come to grips with the very complex philosophy and theory associated with air war at the operational level. This book is for combat officers of any Service who might find themselves on an operational-level staff. More specifically, it is for the air officer who wants to think about air campaigns before called on to command or staff one. It is devoted to how and why air power can be used to attain the military objectives needed to win a war."[18]

Two important ideas are clear in Warden's remarks: first, *The Air Campaign* was for planning at the operational level of war; and second, it was intended to be used by midgrade officers serving on planning staffs who needed to think about how to employ airpower. It was, inside the Department of Defense, something of an intellectual treatise and therefore not exactly accepted in the fighter pilot community (those officers who would actually do the air campaign planning).

Warden and his staff used his ideas to create the Instant Thunder air plan. Warden's Instant Thunder plan wound its way through the approval process and received acceptance and "go-aheads" from Dugan, Loh, and even Schwarzkopf. The only officer involved in the process of preparing for war against Iraq who had not seen combat was the man who actually held the responsibility and authority for executing the plan: Lt. Gen. Charles A. "Chuck" Horner. A Les Misérables confrontation was brewing.

The story told repeatedly and now moving into mythic status of the confrontation between Lieutenant General Horner and Colonel Warden does not bear repeating here. Suffice it to say, Warden and members of his team arrived in Saudi Arabia with the Instant Thunder plan ready to present it to Horner and his senior staff. Warden's briefing of Horner did not go well. Warden was unceremoniously kicked out of Saudi Arabia, but he left behind a strong core of planners to prepare for the air war. The most notable member of this contingent was Lt. Col. Dave Deptula. Deptula and the remnants of the Checkmate team formed the core of officers who made up the Special Planning Group preparing the air war against Iraq, led by Brig. Gen. Buster Glosson. In four months' time, the plan went through numerous revisions and changes, but by early January 1991, it was ready. Taking the preliminary work done by Warden and his Checkmate staff (Instant Thunder), the Special Planning Group took molded it into an executable air campaign that Horner could present to General Schwarzkopf. Years later, a commander of the Air Combat Command remembered that the Air Force was about to get "the Air War we always wanted."[19]

Night One: "Going Downtown" and Gaining Air Superiority

An integrated air defense system is actually a "system of systems" composed of AAA, surface-to-air missiles, and defensive fighter aircraft all linked by a complex structure of early warning radar sites, acquisition radars, fire control and tracking radars, ground control intercept sites, relay stations, and other aircraft and necessary support systems. On paper, Iraq's IADS was top of the line in 1990. As part of this IADS network, the Iraqi Air Force maintained between 500 to 750 combat aircraft, and while most of these were of Soviet design, Iraq's attempts at broadening its inventory led to the acquisition of both French and Chinese aircraft and helicopters as well.[20] The *Gulf War Air Power Survey* (*GWAPS*) later set the number of fighter aircraft at 728, bombers at 14, training aircraft at 400, and helicopters at 511. The SAM threat initially included more than 120 individual batteries, an estimate later raised to more than 200 batteries, as well as thousands of pieces of AAA. This included sites around the capital, as well as scattered throughout Iraq, all connected to the overarching French-designed KARI (Iraq spelled backward in French) integrated air defense system.[21]

"Going downtown" into central Baghdad was a dangerous mission and was reserved, initially, for the pilots of the F-117 Nighthawks, whose low observability offered protection against the Iraqi IADS. On the night of 16–17 January 1991, Iraqi forces unleashed surface-to-air missiles and a curtain of AAA into the sky. Protecting the city of Baghdad that first night were 552 individual SAMs and 1,267 AAA guns, which fired off a "horrific barrage" into the nighttime sky, surprising some of the F-117 pilots who expected to approach the city unnoticed. Stealth pilot Phil McDaniel, in the second wave of stealth fighters that night, turned his F-117 across the heart of Baghdad flying east to west as he headed for an AM radio station followed by an attack on a radio-relay station at Al-Taqaddum Air Base. McDaniel recalled that "the AAA was so thick that the airplane was bouncing with turbulence from all the disturbance in the air." He assessed that the first wave of F-117s had "scared them and kicked up a hornet's nest of AAA." Since the F-117s had no defense armaments and there was nothing in the way of radar warning—the aircraft relied exclusively on its stealth capabilities—McDaniel chose to ignore the aerial barrage occurring outside his aircraft and focused on locating and destroying his targets that night.[22]

All the nonstealth pilots flying that first night and on the nights that followed did not have the luxury of ignoring the AAA, SAM, and MiG threats the way the F-117 pilots did. Saddam's pilots and IADS operators put up a week-long—though largely ineffective—resistance to the coalition air forces. Despite the lack of coordinated effort, this did not mean the coalition enjoyed an "easy" combat experience. In doctrinal parlance, they did not enjoy air superiority right from the beginning; it had to be gained. Clausewitzian fog and friction also remained a potent obstacle to mission accomplishment in the opening hours of the air campaign. Hindsight showed the Iraqi Air Force to be largely inept, but this did not mean they were completely ineffective.

To gain air superiority in those early days, walls of F-15s swept across the eastern and western edges of Iraq looking for targets to kill. In some locations, chaos reigned as pilots attempted to determine friend from foe in the aerial arena. One F-15 pilot, tracking an Iraqi MiG-25 that was in full afterburners (AB), hesitated in taking a shot, fearing fratricide. He opted instead to ask over his intercom, "Is anybody in burners?" When he received terse affirmatives, he ordered, "Everybody out of burners." Afterburners from F-15s blinked out of the Iraqi sky, so when his target continued in AB, he fired a missile and shot down the MiG. It was one of forty-four air-to-air kills during Desert Storm.[23]

Capt. J. L. Briggs, who flew in an F-111F that first night, later pointed out that American training exercisers increased overall effectiveness against Iraqi defenses particularly at night: "Here I was on TFR [terrain following radar], a thousand feet off the ground, going 500 miles an hour, I mean, who does that?" As he approached his targets, he could see the AAA in the distance: "23 mm looks like a fire hose and 57 mm looks like flaming basketballs. It looks impenetrable." As he neared the target, he noted that the AAA seemed to move apart: "As you got closer you realize it's a big sky and you can find a path through the AAA."[24]

The first week of the air portion of the overall campaign to eject Iraq from Kuwait proved to be the heaviest in terms of Iraqi resistance both in Baghdad and over the other defended sites, including the airfields in western Iraq. McDaniel remembered, "AAA was especially heavy the first week—we typically flew every other night and ... planned other missions on our off nights—maybe they were running out of bullets." Briggs noted the same thing: "The first week was the worst and towards the end no one was shooting at us." As the campaign progressed, Briggs and the other F-111F pilots moved from hitting high-value targets to "tank plinking." The Iraqi air defenses had quit responding.[25]

Suppression of Enemy Air Defenses

It is difficult to measure with exactitude how many aircraft the coalition destroyed—both in the air and on the ground—during the conflict. The *GWAPS* report indicated the Iraqi Air Force lost 366 fighters, eight bombers, and 152 trainers. This accounting is complicated by the lack of accurate figures on how many aircraft were flown to Iran and how many aircraft were destroyed while inside hardened aircraft shelters. The bottom line, though, is that Iraq was left with very little in the way of a functioning combat air force in the wake of Desert Storm. Furthermore, Iraq's ability to train its pilots was also significantly degraded.[26]

Of the IADS, a similar story unfolded with more than 100 of the batteries destroyed and only 85 remaining in service after the conclusion of the air campaign. Of the more than 7,000 AAA pieces, the coalition destroyed only about 1,650. Although not within the scope of this work, the Iraqi Navy and Army suffered just as much damage—if not worse—than the Iraqi Air Force. Iraq's missile operators also proved to be poorly trained and ineffective at the employment of their weapons systems during the conflict. However, ineffective does not mean they were fruitless in their efforts. Iraqi surface-to-air missiles shot down more than a dozen aircraft and AAA another five.[27]

The air war continued for six weeks. What once was considered a strategic asset, the B-52, provided tactical airpower against Iraqi forces, in the field, not unlike it had during the Battle of Khe Sanh during the Vietnam War. Tactical fighters eventually went downtown to strike strategic targets. The strategic and the tactical merged. There was no longer "strategic" airpower or "tactical" airpower—there was simply airpower. The coalition, US military, and USAF had it and the Iraqi Air Force did not. It is important to note that despite the overall success of airpower during Desert Storm, this was far from a perfect or clean war. One need look no further than the bombing of the Amiriyah shelter, outside of Baghdad, where over four hundred Iraqi civilians were killed, to see that coalition intelligence proved far from omniscient. Even the ability of the USAF and coalition forces to strike with a level of precision never before witnessed did not mean that all those precision-guided munitions (PGMs) were applied against their intended targets. American planners, although they knew the Amiriyah shelter had been a civil defense bunker during the Iran-Iraq War, believed it had recently been converted to a command center.[28]

What Happened?

So what exactly happened during the six-week air campaign, or rather, the air portion of the overall campaign to eject Iraqi forces from Kuwait? The Desert Storm air campaign seems entirely separate from the dominant doctrinal idea of air-land battle, where airpower works closely to support maneuvering ground forces; however, there were clearly components of this in the ground war, which followed on the coalition's attainment of absolute air supremacy. It seems clear with the benefit of thirty plus years of hindsight that several things occurred during the campaign. First, technology played an important role, and though that could be said of any conflict, in Desert Storm the use of low observable technology and PGMs had an impact not only on the outcome of the conflict, but also on force structure and procurement for the next generation of Air Force development. Second, it seems clear that changes made in both force structure and training in the post-Vietnam years paid dividends in the Gulf War. Third, the Iraqi Air Force (and military writ large) simply did not put up the type of coordinated resistance American military planners thought it was capable of. One could easily get lost in counterfactual arguments of "What would the air war have looked like if Iraq had put up a well-coordinated defense using more of its combat aircraft?" The answer is probably that the Iraqi Air Force would have suffered greater casualties, but then so would the USAF and its allies. The results probably would have been largely the same.

On paper, the Iraqi Air Force looked relatively equivalent in numbers and aircraft "types." Yes, it flew MiG-29s and the USAF flew F-15s, but both also flew Vietnam-era aircraft as well (F-4s and MiG-21s, respectively). The difference in the two sides came largely in the areas of education and training, two arenas where the coalition forces clearly were superior. As noted in the previous chapter, the changes adapted after Vietnam provided massive dividends in the arena of combat.

There might be nothing more emblematic of the role of airpower in the Gulf War than the F-117 Nighthawk. The black jet combined advanced mission planning, the role of technology in aircraft design, and an otherworldly look. When compared to every other fighter or fighter-bomber of its generation, it stands out as truly unique, and this was upheld by the work it contributed during Desert Storm. Not truly a fighter, bomber, or an attack aircraft, the F-117 penetrated undetected into the heart of Iraq to strike strategic targets. Based out of King Khalid Air Base, Saudi Arabia, the 37th Tactical Fighter Wing proved to be, according to Brigadier General Glosson, "the backbone of the strategic air campaign."[29]

Operation Desert Storm confirmed that the inclusion of more training and equipment in the Cold War force structure was indeed prudent. Red Flag proved its usefulness as Iraq suffered the loss of forty-four aircraft in aerial combat. Fighter pilots spoke of their training prior to combat. Historian Williamson Murray said, "Here again peace-time training paid huge dividends. A substantial portion of the air crews, particularly mission and package commanders, had flown in Red Flag." The logistic air bridge, provided by the Air Force's Military Airlift Command, proved not only important but absolutely indispensable. PGMs came into their own during Desert Storm, and their employment would increase and eventually surpass the use of "dumb bombs."

Although they accounted for only about 10 percent of munitions dropped, this foreshadowed what would become an ever-increasing reliance on the technology.[30]

Northern Watch and Southern Watch

After the "end" of the Gulf War, media attention turned to military operations in the Balkans. However, years before the first NATO bomb was dropped in Europe, the Air Force continued to prowl the skies over northern and southern Iraq. The interwar years, in this case between Desert Storm and Iraqi Freedom, were not kind to the Iraqi Air Force and Iraqi defense forces. Operations Northern Watch and Southern Watch not only kept Iraq from conducting aerial attacks against its own people; it also kept the Iraqi Air Force from significant training, and thus allowed American airpower access to continue the "degradation" of what remained of Iraq's air defenses. During this interwar period, the United States routinely engaged both Iraqi aircraft and ground-based defense systems and also fired cruise missiles into Iraq on three separate occasions: in 1993, 1996, and 1998. These actions culminated in 1998 with Operation Desert Fox.[31]

Some Air Force leaders were of the belief that Operation Desert Storm never really ended for the USAF. A letter written in 2012 by Lt. Gen. David Goldfein, the head of Air Force's Central Command, stated that the conflict with Iraq amounted to a continuous operation that had, by that time, lasted more than twenty years. The scale and scope of the operation were significantly less, but members of the Air Force believed that they were left holding the line against Saddam's battered military. While significant combat had ended, the logistics necessary to keep aircraft operating in theater did not get reduced from what it had been during the campaign. The Air Force continued to need aircraft maintenance crews, logistics personnel, and a plethora of other support personnel in the Persian Gulf to conduct daily flying operations. Even as some American pilots were engaged in the Balkans, others deployed to the Middle East to contain what was left of Saddam's forces. These rotations to the Middle East to keep Iraqi forces boxed in contained Saddam's forces in the country's interior, but the deployment was an additional chore for American pilots that they generally did not like. US pilots split their time in the 1990s. While at home station they performed continuation training and attended Red Flag and other live-fire exercises, including Combat Hammer and Combat Archer, where the munitions leaving the rails were real. When not honing their skills in the United States, the fighter squadrons and support aircraft rotated through CENTCOM. Beginning with the cease-fire on 3 March 1991, Desert Storm soon gave way to the northern and southern no-fly zones (NFZs). The NFZs, for the most part, proved to be little more than opportunities for pilots to "drill holes in the sky," as the parlance went. The operations did provide combat experience for any pilot who flew when Iraq ill-advisedly attempted to violate the NFZs. After the war ended, a pilot faced more of a "threat" at a Red Flag than he had faced from what remained of Saddam's air defenses.[32]

The no-fly zones had been initiated after Iraqi helicopters began using chemical weapons against forces opposed to Saddam's regime in southern Iraq in March 1991. Furthermore, Iraqi fighters began flying sorties as well. In response, General Schwarzkopf ordered Air Force fighters back into Iraqi airspace to ground the Iraqi Air Force.

Not only had Iraq's IADS been knocked out during Desert Storm, but the Iraqi military was unable to reconstruct it after 1991 because American airpower continued to box in the country as part of Northern and Southern Watch. Set up to prevent the Iraqi dictator from conducting further reprisals against his own people, the Northern and Southern Watch operations prevented Saddam from reestablishing a countrywide IADS. This also led to the claim by the US Air Force (somewhat erroneously) that it never left Iraq after Desert Storm. It became a point of pride, which then–lieutenant general David Goldfein mentioned in his 2012 article.[33]

These "look down–shoot down" engagements began on 20 March 1991, when an F-15C shot down an Iraqi fighter aircraft. Two days later, an F-15C shot down an Iraqi SU-22 near Kirkuk in northern Iraq. On the same day, another USAF flier literally intimidated an Iraqi pilot into ejecting from his aircraft shortly after taking off rather than engage with the American. These incidents proved to be the last time Iraq attempted to launch aggressive aircraft for the next year.[34]

These early signs of Iraqi desire to be more aggressive proved to American airmen that combat operations, at least for the Air Force, had not ended with the cease-fire. Throughout the rest of 1991 and into 1992, Air Force fighter aircraft continued to patrol both northern and southern Iraq. In response to the Iraqi government's continued use of helicopter-borne weapons against civilians, the coalition members instituted a complete no-fly zone in Iraq beginning in the summer of 1992. In December 1992, Lt. Col. Gary North garnered a number of firsts when he shot down an Iraqi MiG-25. It was the first F-16 air-to-air kill and the first use of the AIM-120 as a beyond-visual-range kill.[35]

Despite the success of realistic training exercises back in the United States, one incident demonstrated that both training and technology could fail with disastrous results. This incident resulted in the death of twenty-six military members. On 14 April 1994, two F-15s incorrectly identified two UH-60 Black Hawks as Russian Mi-24 Hinds. American E-3 AWACS confirmed the helicopters as bandits, and the F-15s fired two AIM-120 missiles. Both helicopters were destroyed. The AWACS controllers failed to notice that their scopes showed the helicopters as friendly, and the two F-15 pilots misidentified the helicopters during a visual inspection fly-by. Former Army officer and Harvard professor Scott A. Snook noted in his book *Friendly Fire: The Accidental Shootdown of U.S. Black Hawks over Northern Iraq* (2000) that both F-15 pilots were "highly trained, technically qualified, and well-respected officers with hundreds of hours experience in the aircraft." Likewise, the AWACS controllers were "trained and equipped to track literally hundreds of enemy and friendly aircraft during a high-intensity conflict." The postincident investigating board looked into the "training and readiness programs" in which the pilots had participated. The board's final report did not blame technology; rather, it stated that "neither F-15 pilot had received recent, adequate visual recognition training." It was the human element that failed in this instance.[36]

Over the next decade, American forces continued to patrol the Iraqi northern and southern NFZs. These operations often turned hot when Iraq would launch aircraft or turn on surface-to-air missile radars, and US forces responded by destroying the Iraqi weapons systems. Operations in Iraqi no-fly zones proved useful to the USAF because they allowed American airmen to gain combat experience in a low-threat environment, but they did not prove as effective in honing needed combat skills as participation in large-force employment training exercises back home in the United States.

The utility of the NFZs proved sound for the contextual situation faced inside the borders of Iraq. As a way of war, for a military, they more closely resemble a siege or blockade and, as mentioned, were not liked by the pilots who had to fly them. However, the *idea* of a no-fly zone proved to be appealing to policymakers. As a tool, governments and administrations enjoyed the ability to employ a no-fly zone. It allowed something to be done, but each contextually different situation begged the question of just *whom* you wanted to prevent from flying.

Air Campaigns in the Balkans

The USAF sometimes joined by the air forces of other NATO allies conducted Operations Deny Flight, Deliberate Force, and Allied Force over the Balkan Peninsula in the 1990s. The Air Force entered the conflicts in the Balkans with an overinflated sense of itself and an identity that, being now firmly aligned with tactical airpower, could not, at the end of the day, conclusively win a conflict in the absence of contributions from other services. Airpower did a lot during the progressive campaigns, but airpower alone did not win in the Balkan conflicts. This is actually somewhat of a controversial statement. No less a historian than the eminent John Keegan boldly stated, "Now there is a new turning point to fix on the calendar: June 3, 1999, when the capitulation of President Milosevic proved that a war can be won by air power alone." This is entirely debatable. A number of contextual elements indicate that airpower alone did not bring Operation Allied Force to a successful conclusion. Only through diplomacy, the threat of further force to include large numbers of NATO ground troops, and the withdrawal of Russian support for Yugoslavia did the Balkan conflict come to an uneasy truce, if not a conclusion. That did not stop some from heralding the victory in the Kosovo War as a victory for airpower.[37]

Three independent and yet closely related air campaigns represent the bulk of the combat in the Balkans for the USAF: Operations Deny Flight, Deliberate Force, and Allied Force (Noble Anvil). It is necessary to look at each conflict individually to determine what the USAF did or did not accomplish in these engagements. To be sure, the Balkans campaigns were complicated affairs. Entire books could be written about the causes, conduct, and outcomes of each of these operations. Suffice it to say, the focus here is primarily on the actions of the US Air Force, and to a lesser extent on the rest of the US aerial forces and coalition air forces as well. For reasons difficult to explain, the varied air forces of the United States found themselves asked to conduct operations in the skies over the former Republic of Yugoslavia.

Operation Deny Flight

Deny Flight was a NATO operation that ran from April 1993 through December 1995. As its name implies, it was an enforcement of a no-fly zone, authorized by the United Nations, over Bosnia and Herzegovina. Although the operation proved successful at keeping fixed-wing fighters out of the NFZ, rotary-wing assets proved to be more difficult to intercept.

The most notable event during Deny Flight was what came to be known as the Banja Luka incident, which occurred on 28 February 1994. This proved to be the first hostile shots fired by NATO in its history, and these first shots came from the wings of USAF F-16s. The incident involved eight aircraft belonging to the Republika Srpska Air Force: six J-21 Jastrebs and two J-22 Oraos. After violating the NFZ and bombing a munitions plant in the city of Novi Travnik, they were engaged by four Air Force F-16s armed with AIM-120 AMRAAMs and AIM-9 Sidewinders. Capt. Robert "Wilbur" Wright, flying lead in the first pair of F-16s, sent his wingman, Capt. Scott O'Grady, into "high cover" and shadowed two of J-21s from below and behind. Wright informed the aircraft they were in violation of UN Security Council Resolution 816, but he received no response. Four more aircraft popped up (the other two J-21s and two J-22s). Wright was now faced with a 2v6 engagement and later admitted to feeling a bit "offensive." Wright called in a second pair of F-16s to even the odds. The original two J-21s joined the other four aircraft and all rolled in on their bombing run. Having committed a hostile act, they were now legitimate targets for Wright to engage. After coming off their target, all six enemy aircraft were now in front of and being trailed by the four American F-16s. Wright, being in the best offensive position, fired first. He fired three missiles for three kills, an unusually high Pk rate, as all of his missiles (one AIM-120 and two AIM-9s) found their mark. O'Grady engaged next, but his AIM-9 was out of range and missed. Wright and O'Grady pulled off, and the second set of F-16s engaged the remaining three aircraft for another kill. The ending tally was four enemy aircraft destroyed. Ironically, Wright and O'Grady would be involved in another incident fifteen months later, this one with a different outcome.[38]

On 2 June 1995, American aircraft were continuing to patrol the no-fly zones. Capt. Scott O'Grady (Basher-51) was part of a "two ship" flying as Wright's (Basher-52's) wingman when he was targeted by a road-mobile SA-6. The radar and missile operators were turning their systems on intermittently. This caused a disconnect between Wright, O'Grady, and their E-3 AWACS. Wright picked up the radar emissions first, calling, "Basher-51, mud six, bearing zero-nine-zero." Wright was informing both O'Grady and the AWACS that he was picking up a Yugoslav radar. AWACS reported back that this was an "uncorrelated" radar track. O'Grady's F-16 was also not showing an enemy radar. Moments later, O'Grady received the same warning indicator Wright had received previously. Two missiles were in the air and on their way toward the pair of F-16s. The first passed between the two aircraft; the second blew O'Grady's aircraft in two.[39]

Captain O'Grady successfully ejected, but he would spend the next six days evading enemy forces on the ground. After O'Grady's rescue, there was no shortage of finger-pointing and blame-laying, some at the feet of O'Grady himself. To complicate matters, while O'Grady was on the ground, Chief of Staff of the Air Force Gen. Ronald Fogleman announced to reporters that "intermittent" transmissions were being received from O'Grady's rescue radio. A NATO official stated, "I was dumbfounded he said that . . . I mean, why not just announce to the bad guys, 'We think he's alive and kicking, and we hope we find him before you do'?" Despite this, O'Grady was rescued successfully on 8 June.[40]

Deny Flight also involved occasional attacks in support of UN forces on the ground. These included air strikes in August, October, and November 1994. In the end, Deny Flight proved relatively successful. The operation ended in December 1995. Beyond the two incidents above, Deny Flight was largely uneventful and left little room for objective review as to whether or not airpower was successful in the Balkans. The next two operations provided much more material to study.

Operation Deliberate Force (ODF)

Deliberate Force was a combined United Nations–North Atlantic Treaty Organization "dual key" military operation, meaning that leaders from both organizations needed to grant permission for military action to begin. Permission was gained and offensive operations began the night of 29 August 1995 and lasted through 20 September 1995. The plan called for air strikes against the Bosnian Serb Army (BSA), including on lines of communication, the IADS, and other military targets. In this vein, the operation was largely successful and helped the warring parties to achieve the resultant Dayton Agreement, signed in November 1995.

Most every conflict lays claim to notable "firsts," and two aspects of Deliberate Force bear mentioning as they were certainly harbingers of things to come for the United States Air Force. ODF was the first conflict where PGMs outnumbered unguided munitions ("dumb bombs"). This meant that for the first time, the Air Force's preferred method of weapons employment transitioned from precision delivery to precision guidance. Heralding things to come, ODF also saw the first large-scale employment—as well as shoot downs—of the Predator unmanned aerial vehicles (UAVs). When coupled together, the first use of the Predator and the maturation and delivery of PGMs demonstrated a significant technological shift for the USAF. For a service that focused on technology, ODF represented a significant, positive change in the Air Force's way of war, but the effects and conduct of the war left a bit more to be desired.[41]

The main threat facing coalition aircraft was not so much the BSA's air force, but its IADS, which included SA-2s, SA-6s, and SA-9s. It represented a potent threat and was therefore of primary concern during offensive operations. More of a hindrance, though, was the weather. It was rapidly recognized that this was not Iraq, and that the weather had a say in just when and how often combat sorties were going to fly and whether an attack mission could even locate its target.

Col. Christopher M. Campbell stated in his chapter "The Deliberate Force Air Campaign," "One doesn't drop bombs on target sets of COGs [centers of gravity] but on things—hopefully things the adversary considers valuable." Thus, it must be asked what target sets were attacked by the Air Force and other coalition air forces. The target matrix that was used largely followed the successful planning of Operation Desert Storm and included air defense/IADS, command and control, essential military facilities, lines of communication, and infrastructure—in essence, anything that allowed the BSA to function as an effective fighting force.[42]

There was another unusual aspect of Deliberate Force that bears noting and proved to be unlike anything the Air Force had done in the past. Much of the offensive airpower flew out of Aviano Air Base in Italy. Pilots permanently stationed there were on three- to four-year standard "accompanied" tours, meaning their families were with them. Unlike Desert Storm, or Vietnam, these pilots left home each day, flew their combat missions, and returned to hearth and home. This arrangement proved to be both a benefit and a hindrance. On the one hand, as one officer noted, "officers at Aviano appreciated having their families with them to provide support. Spouses provided meals in the squadrons every evening as a way of supporting the unit and providing comfort to one another." On the other hand, it was interpreted by some as a distraction to have the families colocated in what was essentially a base conducting active combat operations. One squadron commander believed that in the event the conflict continued, "spouse or families should have been moved out of the area to eliminate a potential distraction for his crews."[43]

Operation Allied Force

Three days. That was initially how long air force planners believed the strategic aerial attacks against Yugoslavia would take to force a capitulation and cessation of hostilities from Slobodan Milošević. In short, NATO wanted Milošević's Yugoslav forces to cease their ethnic cleansing operations against the semiautonomous region of Kosovo. Following the 1995 NATO-led Deliberate Force operation, Allied Force marked the second use of military might by NATO, and the first accomplished without the direct support of the United Nations Security Council. The strictly American portion of the operation went by the name Noble Anvil. Allied Force seemed an ideal setup for the use of airpower—and by extension, the nonuse of "boots on the ground." If NATO leaders and US president Bill Clinton wanted to keep from entangling forces in a very messy situation on the ground, but at the same time to apply pressure against Milošević, then it seemed airpower provided the mechanism to achieve that. Since Desert Storm, airpower had proved to be an ideal tool for conducting military operations without the associated risk of placing American soldiers or Marines in harm's way. Aerial attacks began on 24 March 1999 when NATO Secretary-General Javier Solona "directed SACEUR [Supreme Allied Commander Europe], General Clark, to initiate air operations in the Federal Republic of Yugoslavia." Rather than last just three days, the campaign dragged on for seventy-eight days. Disputes and disagreements were commonplace between Gen. Wesley K. Clark and his air component lead Lt. Gen. Michael Short. As the campaign dragged on, the air planners and NATO staff steadily ratcheted up the pressure through ever-increasing aerial attacks.[44]

Again, a potent threat of SAMs that were part of a robust IADS posed problems for air planners. The missiles—of the SA-2, SA-3, and SA-6 varieties—

themselves numbered in the thousands. These were backed up by an extensive set of radars and ground observers all inhabiting a relatively small area that was itself hemmed in by mountains and prone to significant periods of cloud cover and extreme weather not exactly conducive to air operations.[45]

On the first night of operations, five MiG-29s lifted off to counter the allied armada. They faced dozens of USAF F-15s. The 5vX battle ended with three of the MiGs' shot down. Lt. Col. Cesar Rodriguez downed one of the MiGs that night, bringing his aerial victory credits to three. This made Rodriguez one of only a few pilots to achieve that feat since the end of the Vietnam conflict, one of the others being Captain Wright during the Banja Luka incident. That same night, Capt. Michael K. Shower was flying as part of a combat patrol protecting F-117s and B-2s in the initial strike package. Shower located a MiG-29 and fired three AIM-120s at the enemy aircraft. The first two missiles, unbeknownst to Shower, flew within 2,000 feet of one of the F-117s. The third missile killed the MiG.[46]

The next afternoon, Capt. Jeff Hwang claimed another two MiG-29s. The final aerial kill came on 9 April when Lt. Col. Michael H. Geczy, flying an F-16CJ downed another MiG-29, bringing the air-to-air kills in Allied Force to a 5–0 tally in favor of the Americans. Clearly, the US technological and training edge continued to pay dividends. Still, the aerial combat over Yugoslavia was not without losses. The threat of surface-to-air missiles remained potent and shocked the USAF three nights into the campaign.

The unthinkable happened on 27 March 1999, just three days into the campaign. Members of the 3rd Battalion of the 250th Missile Brigade, under the command of Col. Zoltán Dani. equipped with SA-3s, the battalion successfully tracked, targeted, fired upon, and shot down an F-117 Nighthawk—firing three, if not more, missiles. As the pilot later recalled, "The first missile went right over the top of me. So close, actually, that I was surprised it didn't proximity fuse on me. I could feel the shock wave of it buffeting the aircraft. As soon as it went over I quickly re-acquired the second missile visually and when I did, I thought: 'It's going to run right into me.' And it sure felt like it did." The missile strike blew the wing off the aircraft sending it into a violent roll that pushed the pilot up and out of his seat. By his own recollection, it took a significant amount of effort to reach the ejection handles.[47]

The pilot of the aircraft, later identified as Lt. Col. Dale Zelko, contacted forces in the area immediately after ejecting from the aircraft as he was still slowly heading for the ground under the canopy of his parachute. He called Mayday three times, and upon receiving an acknowledgment from a nearby E-3 AWACS, he said, "Roger, out of the aircraft . . . downed!" A combat SAR package orbiting nearby heard his initial call "on guard," and the rescue operation began in earnest. In his MH-53, Capt. Jim Cardoso would lead the rescue effort to go in and get Zelko. He knew, as did every other member of the rescue team, that the Yugoslav forces were going to actively attempt to capture the downed pilot before he could be rescued. It was a daunting task, to say the least, but one the crew was well trained for. Cardoso remembers thinking, "A stealth just got shot down and now [they] want us to go in there?"[48]

The route into the rescue area was hazardous. The helicopters needed to fly dangerously close to Belgrade, an obviously heavily defended area bristling with SAMs. Approaching Zelko's position, the rescue team actually flew over three Yugoslav trucks with dismounted troops who were themselves looking for the downed American. Having trouble locating Zelko by both infrared strobe and laser, they radioed him to fire a flare. This helped the pilots find him in the wind, rain, and fog but also gave the Yugoslav team a location on which to concentrate their efforts to capture the American. Setting down next to Zelko, a pararescue jumper grabbed him and hauled him into the MH-53. It took less than forty-five seconds to find Zelko, pick him up, and depart the area.[49]

The air war bogged down with visions of Rolling Thunder dancing in the heads of air planners. A RAND report later concluded, "NATO air attacks continued to be hampered by bad weather, enemy dispersal tactics, and air defenses that were proving to be far more robust than expected." Even more problematic, the situation on the ground with regard to Serbs' ethnic cleansing of Kosovars was getting worse and not better. General Clark's answer to this was increased pressure from the air, which itself necessitated more aircraft. This involved a request for AH-64 Apache attack helicopters that proved controversial because of the Apaches' association with the support of ground troops. Throughout the campaign, the number of aircraft assigned for support continued to rise, from around 400 aircraft at the inception of combat operations to over a thousand at the cessation of hostilities. This number of combat aircraft in fact exceeded that used during Desert Storm.[50]

Targets attacked in the first week of the campaign were retargeted and struck a second time. In some cases, these targets needed to be struck again and again. In part this was due to Clark's "desire to keep the bombs falling," and partly it resulted from Yugoslav forces' ability to repair or replace damaged equipment. The Yugoslav forces proved to be a competent foe and, unlike the Iraqis in 1991, were able to keep their IADS a functioning threat throughout the Allied Force campaign.[51]

On 2 May, two months after the shoot down of the F-117, the squadron commander of the 555th Fighter Squadron was also engaged and shot down by SAMs. The pilot, Lt. Col. David Goldfein, was rescued as part of another daring CSAR mission. After having the rear of his aircraft shredded by a SAM,

Goldfein reported that his F-16CJ was now "a glider." Moments later and while descending in his broken aircraft, Goldfein told his fellow pilots, "Start finding me boys." The other F-16s remained overheard to provide on-scene command and control of the rescue operation. As in the rescue of Lieutenant Colonel Zelko, the rescue helicopters approached, but this time they were able to see the downed pilot's strobe light. They quickly dropped into position and a para-rescue team disembarked to pick Goldfein up, close to simultaneously with the approach of a Yugoslav team sent to capture the downed pilot.[52]

A mere five days after the downing in Goldfein, on the night of 7 May, an American B-2 Spirit dropped a series of precision-guided joint direct attack munitions (JDAMs) on the Chinese Embassy in Belgrade. Using faulty data—there remains dissension on just how the incorrect building was targeted that night—the JDAMs hit the embassy close to midnight. Three Chinese nationals were killed in the attack, and another twenty were injured. All investigations into the bombing proved the attack was accidental and not a deliberate act. This did nothing to counter the perception that Allied Force was not working.

NATO planners were eventually forced to ratchet up their aerial attacks yet again, this time against Yugoslavia's electrical system. These attacks came dangerously close to reigniting the debates about strategic bombing in World War II. For although the attacks were ostensibly targeted against the Yugoslav military forces, they also placed a disproportionate amount of pain on the civilian population of Yugoslavia. Millions were left without power.

One further story bears telling here that involves the limits of technology and violation of the Air Force rule of "centralized control and decentralized execution." The essence of this principle is that planners plan and then give pilots leeway to execute the mission without prohibitive interference. For example, during one mission an A-10 pilot was having difficulty locating a tank. A Predator was relaying a live video feed of the tank directly into the Combined Air Operations Center (CAOC), but the pilot, flying thousands of feet above the drone, had a significantly different point of view and simply could not see the tank, although everyone on the floor of the CAOC could. Lieutenant General Short, becoming agitated, had a message relayed to the pilot that he personally wanted the tank destroyed. The A-10 pilot curtly replied to the controller inside the CAOC, "Tell *Dad*, I can't find the fucking tank."[53]

Operation Allied Force came to a close on 10 June 1999, having lasted some eleven weeks longer than initially anticipated. Even with twenty years of hindsight, the legacy of Allied Force has yet to be determined. Truly, many "firsts" occurred during Allied Force/Noble Anvil: the first large-scale use of unmanned aerial vehicles for ISR purposes. Unknown or unrecognized at the time, UAVs were soon to become ubiquitous in war, in and from the air. The first combat use of the B-2 also occurred with flights departing and returning to Whitman Air Force Base, Missouri. Perhaps most importantly, no NATO aircrews were killed or taken prisoner. The limits of airpower were in evidence again as technology showed its vulnerabilities in the downing of the F-117, and despite ever more technologically advanced aircraft and avionics, weather still played a role in warfare as well. The potent IADS threat demonstrated how it could be used effectively to counter American aerial supremacy, although losses were less than anticipated.[54]

Retired USAF Lt. Col. William O'Conner wrote in his memoir *Stealth Fighter* that, following his participation in Allied Force, he returned home, thinking: "So after becoming a combat veteran and witness to history in a little odd conflict, I waited for the stories to come forth so that I could read about *my* war. With few exceptions, I'm still waiting." There are several reasons for this, not the least of which is that the proximity to the conflict means that many of the relevant primary sources, from mission reports to air attack plans, remain classified. With such sources still being within the twenty-five years most documents are required to remain classified, it is yet to be seen if the smaller conflicts of the mid-to-late 1990s will garner a resurgence of interest when they become available to researchers. Despite this, there already exist enough sources to discuss what airpower did and did not do during Operation Allied Force.[55]

Conclusion

The 1990s proved to be a golden age for the United States Air Force, despite the loss of the Soviet Union as the nation's principal opponent. Desert Storm and, to a lesser extent, the Balkans campaigns seemed to prove the efficacy of airpower as both a policy and military tool, and militarily it worked exceptionally well against the enemy it was asked to confront. Airpower became the go-to instrument used by the Clinton administration in order to force or compel an enemy toward a desired outcome. While the merits and success of the Balkans operations continue to be debated, it seems, in retrospect, that as a policy tool airpower was largely effective, if not always as instantly as desired, in what it was asked to do in the 1990s.

Technological advancements of the 1970s through the 1990s became normalized in Air Force operations, including precision-guided munitions and low observable aircraft. Despite the loss of the F-117 over Kosovo, the Air Force could still point to the utility of stealth aircraft in attacking targets in deeply defended terrain. Aircraft losses dropped significantly since the days of Vietnam, and new aircraft coupled with advancements in training gave every indication that airpower maturation had finally reached the hopes and dreams of the likes of

Mitchell and de Seversky. Operation Desert Storm was a resounding success by any military measure, and Allied Force was heralded as a victory through airpower alone. However, the coming of the new century brought with it the revival of an old style of warfare, one in which the Air Force struggled to find its place.

The Air Force now had a clearly defined "way of war" using technologically advanced aircraft, precision-guided munitions, and exceptionally educated and trained aircrews capable of engaging with and destroying any peer or near-peer competitor. However, it was not imagined that the next enemy might not be a peer nation... or a nation at all. The wheel turned and the Air Force again found itself facing an enemy that did not conform to its preferred method of combat.

Procurement of aircraft and other weapons systems continued even as the twentieth century gave way to the twenty-first. In December 2005, the F-22 reached its initial operational capability. It would be followed a decade later by the F-35, but both of these aircraft were purchased in significantly fewer numbers than the aircraft they were "replacing," such as the F-15 and F-16. As the twenty-first century began, there emerged valid objections to procuring a fleet of aircraft that were practically useless against the enemy that reared its head on 11 September 2001 with attacks that would initiate "the forever wars."

CHAPTER 8

Operations Enduring Freedom and Iraqi Freedom

Oh, East is East, and West is West, and never the twain shall meet,
Till Earth and Sky stand presently at God's great Judgment Seat

—Rudyard Kipling, "The Ballad of East and West," 1889

It took several years for the comparisons between the war in Afghanistan (Operation Enduring Freedom) and the conflict in Vietnam to arise, but after more than a decade of fighting—ultimately, some twenty years—the comparisons were inevitable. The "liberation" of Iraq from Saddam Hussein's regime (Operation Iraqi Freedom) lasted from 2003 to 2011, but in 2014 the USAF again found itself attacking targets in western Iraq to counter the rise of the Islamic State. Two decades of continual combat in Afghanistan and another sixteen in Iraq (and now Syria) left USAF leaders with more questions than answers, but this should not be the case. Airpower has shown a unique ability to provide utility across a wide spectrum whenever and wherever it has been needed. Rather than struggle with what it is has provided, the USAF should recognize it provided whatever was needed, wherever and whenever it was needed—whether that entailed close air support, interdiction, or strategic attack. The Air Force has demonstrated the unique ability to fit its assets to any task it is called upon to perform.

If the years 1991–99 represented a preferred way of war and what has come after the 9/11 attacks represents an inchoate war against an asymmetric enemy, this transition did not indicate a need to change tactics and doctrine for Air Force leaders in the twenty-first century. What has happened between 2001 and the present does not represent either a lost way of war or a loss of airpower theory. Truly, although Afghanistan could be considered a style of war that did not conform to existing theory, that does not mean that the USAF was incapable of using its assets well in this asymmetric fight. The contrary proved to be true. Airpower was pivotal in the war in Afghanistan, perhaps more decisive even than in any previous campaigns.

When States Do Not Fight

Ironically, just as the Air Force proved the full dominance of air operations against peer and near-peer competitors during the 1990s, it was asked to provide combat airpower against nonstate actors at the dawn of a new century. The 9/11 attacks led to Operation Enduring Freedom in Afghanistan, and despite al-Qaeda's and the Taliban's lack of an air force, the USAF proved indispensable in the early campaigns against both organizations. A combination of Special Forces and airpower overwhelmed the enemy forces and effectively removed the two groups from leading the country or finding sanctuary in it within a matter of weeks. So precise and overpowering were the air strikes that Secretary of Defense Donald Rumsfeld quipped, "We're not running out of targets, Afghanistan is." However, the early and overwhelming success of Enduring Freedom gave way to a conflict of insurgency, counterinsurgency, and irregular warfare that lasted the better part of two decades. This did not mean that the USAF did not have a role in the continuing campaigns; rather, through the first two decades of the new century, the Air Force provided on-call close air support. Technological advances led to increased use of unmanned aerial systems improved intelligence, surveillance, and reconnaissance capabilities. As the decade wore on, the war morphed into something new as USAF assets combined with the Central Intelligence Agency to wage a new style of warfare: the drone war, where UAVs provided offensive missions. However, the USAF should be hesitant to believe that this represents a paradigm shift for future operations. The geography and air supremacy enjoyed over Afghanistan and Pakistan will not necessarily be the case in future operations.

Operation Enduring Freedom

Less than one month after the terrorist attacks of 11 theSeptember 2001, the United States military launched Operation Enduring Freedom (OEF). The reprisal attacks began on 7 October and targeted the al-Qaeda network and its Taliban supporters in Afghanistan. Initial strikes came from USAF B-52, B-1, and B-2 bombers flying from either their home stations in the United States or from Guam and later Al Udeid Air Base, Qatar. Tactical support and other attack sorties came from the US Navy. This mission required Navy attack and fighter-bomber aircraft to fly long distances with multiple air-to-air refuelings to reach their targets in landlocked Afghanistan. The operation began with limited planning since no major contingency plan for Afghanistan existed. Although Air Force bombers and Navy strike aircraft could reach Afghanistan, it was still necessary to have Air Force search and rescue assets close enough to respond to a downed aircrew situation should something go wrong. These did not arrive in Uzbekistan and Pakistan until a week before hostilities commenced. It was even later, not until 17 October, that USAF tactical aircraft—F-15E Strike Eagles and F-16s—entered the fray, flying from bases in Kuwait.[1]

Early airpower during OEF is best remembered for the use of special operations forces (SOF) troops on the ground, working in conjunction with the Afghan Northern Alliance, to coordinate air strikes against enemy targets. This represented a fundamentally different approach to airpower, but one that in this particular context worked. As American SOF and Northern Alliance forces moved across Afghanistan, they marked targets and a blanket of airpower provided coverage and on-call destruction of these targets. From October through December 2001, SOF and airpower degraded Taliban and al-Qaeda forces. This culminated in the Battle of Tora Bora, where remnants of the enemy forces were either killed in air strikes or slipped through the porous border with Pakistan. It was a culminating battle, but not a decisive one. While the cooperation between SOF and airpower was nothing short of spectacular, it did not constitute a doctrine capable of being sustained.

The working relationship between SOF and airpower also proved to be short-lived. As more regular Army soldiers moved into Afghanistan, the early interservice cooperation evaporated. This was not the result of any form of rivalry, but occurred mainly through the blossoming of the respective staffs: coordination began to fall through the cracks of the staff system. Clearly, the most notable instance occurred during Operation Anaconda. Large numbers of conventional Army troops were engaged with the enemy for the first time, but coordinated air support was not always available or capable of delivering the desired results. In the aftermath of Anaconda, senior leaders in the Army and Air Force pointed fingers at one another. Another compounding factor was that the senior air planners in the region and a portion of their staff were not working in the CAOC but were instead touring other sites in the Middle East and laying the groundwork for an invasion of Iraq. Following Anaconda, operations in Afghanistan took a back seat to the eventual invasion of Iraq in February 2003. Even so, Air Force units continued to supply air support to "troops in contact" for much of the following fifteen years.[2]

It was not until 2011 that the leader of al-Qaeda was finally located. Although a strike composed of US Army aviation assets and the Navy's special warfare unit eventually dispatched the mastermind of 9/11, this was not the only option. For a time, it seemed there might be a plan to bomb the compound where Osama bin Laden had been found in hiding in Pakistan. This attack could have been carried out by any number of aircraft, but within Pakistani airspace the low observable F-22 and B-2 made the most sense. Such an operation was eventually ruled out, and an air strike was dropped from consideration. At any rate, the death of bin Laden did not usher in the end of American involvement in Afghanistan. Nearly twenty years after it began, Operation Enduring Freedom finally concluded. The USAF's presence in the war had remained largely unchanged since the collapse of the Taliban and al-Qaeda in the fall of 2001 until 2021 when the last US forces departed the country and the capital city of Kabul fell in August providing an end to America's longest war. Still, the Kabul airlift conducted by the United States and other allied nations flew more than 120,000 refugees out of the country, with the United States evacuating over 80,000 of these. Years from now when the histories of the American conflict in Afghanistan are written, the Kabul airlift may stand alongside "flying the Hump," the Berlin Airlift, and Operations Nickel Grass and Desert Shield.[3]

Operation Iraqi Freedom

For decades to come, the decision of the George W. Bush administration to invade Iraq will continue to be questioned. Much like the war in Afghanistan, the "rightness" or "wrongness" of the rationale for the invasion will be ignored herein; rather, focus will be given to the actions of the US armed forces in engaging the Iraqis. Eighteen months after the attacks of 11 September 2001, the US military launched an invasion of Iraq, tenuously linked to 9/11 by the Bush administration and with the explicit aim of removing the Saddam Hussein regime from power. On the other side of the battlefield, Iraq's military continued to be hindered from responding to the invasion in almost every conceivable way by a leader who believed a coup was a more likely threat. According to an analysis by Joint Forces Command, to Saddam, "the idea that the Americans would attack all the way to Baghdad appeared ludicrous."[4]

The beginning of the Iraq War was a drastically different story on the opposing sides of the conflict. Twelve years of being boxed in by Operations Northern Watch and Southern Watch had seriously hindered Iraq's ability to respond militarily. The enemy faced by the allied forces was still significantly degraded from the fight a dozen years previous and the engagements that followed, which meant that Iraq posed significantly less of a military threat than it had in 1991. One Iraqi leader offered this perspective: "We thought this war would be like the last one in 1991, we figured that the United States would conduct some operations in the south and then go home." The Iraqi "strategy" was to not fight back. Saddam no longer trusted his air force to respond, so he ordered it to stay grounded; instead, he ordered that aircraft be dispersed away from airfields or concealed away from the perceived omnipresent eyes of American fliers. In at least one instance, this meant literally burying the aircraft in the desert. This bears repeating: during Operation Iraqi Freedom, the Iraqi Air Force did not launch a single sortie against coalition airpower. Iraqi pilots did not participate in the defense of their country against the invaders.[5]

The image of an Iraqi MiG-25 buried in the desert sands became one of the iconic photographs of the 2003 invasion. One of the unanswered questions from 2003 is why did the Iraqi Air Force bury that MiG-25. Conventional wisdom has it—or at least the story told by the National Museum of the United States Air Force—that Iraqi forces buried the MiG-25 to prevent its destruction by coalition airpower in response to leaflet drops telling the Iraqis to "ground" their aircraft. The Iraqi military mistranslated the leaflets and instead of keeping their aircraft "grounded," they "buried" them instead. It was a nice story, but not entirely true.[6] The real answer lay in Saddam's prewar directions that the Iraqi Air Force was not to participate in the war. Saddam ordered his air force to camouflage its aircraft in palm groves or in the desert. In the case of the latter, air force members simply buried nonflying aircraft in the sand. In the case of the MiG, this seems odd considering Iraqi forces had already removed the aircraft's wings. In an age where "effects-based operations" were still discussed, the effect the coalition wanted was achieved regardless of the plane's burial. It was not going to fly anyway.[7]

Thus, for American airmen, the single greatest threat, as before, came from Iraqi SAMs, this time launched autonomously and without the support of their radars. The first night of both Desert Storm and Iraqi Freedom said much about both operations, but in ways the USAF did not expect.

Southern Focus

Historian Benjamin Lambeth titled his book on Operation Iraqi Freedom *The Unseen War,* and this is an especially apt insight, considering coalition airpower executed a war before the war: Operation Southern Focus. Histories of Iraqi Freedom indicate that the air war and ground war occurred concurrently. This is not accurate. Operation Southern Focus began in June 2002, nine months prior to the invasion. Its purpose was to increase the number of kinetic strikes against Iraqi air defense sites in preparation for a possible invasion of Iraq. Targets included not only SAM sites, but early warning radar and command and control nodes as well. Interestingly enough, Southern Focus was not announced to the American public despite coalition aircraft dropping more than 600 bombs between June 2002 and the "beginning" of hostilities in March 2003.[8]

Southern Focus worked remarkably well as it rendered a large portion of the country unable to be defended by Iraqi air defense systems. Coupled with an air force that could not or would not participate, Southern Focus gave allied fliers virtual aerial supremacy throughout much of Iraq before the official start of Operation Iraqi Freedom. The Iraqis moved what remained of their IADS south of the 36th parallel and north of the 33rd parallel into a concentrated area between Baghdad and Tikrit. This area became known as the super missile employment zone, or "Super MEZ."

So what did the Iraqi military of 2003 look like compared to that of 1990, and what type of threat did the country pose to American and coalition forces? An answer is difficult to flesh out because many of the records for Iraqi Freedom remain classified. However, a US Air Forces Central report conducted by the Assessment and Analysis division, released in April 2003, noted that "overall operational capability of Iraqi aviation was low while surface-to-air threat was assessed as medium to high." Apparently, coalition forces had no idea that thanks to "dysfunctional regime policies," the Iraqi Air Force was destined to not launch a single sortie.[9]

The biggest concern prior to the invasion was not the systems that coalition forces knew existed, but rather the possibility that there existed scores of unlocated AAA and SAM sites throughout the country, but especially in the "Super MEZ" area in central Iraq. Once Northern Watch and Southern Watch went into effect in the 1990s and with its defensive posture exacerbated by Southern Focus, the Iraqi military husbanded as many of air defense systems as it could into areas where the Americans did not fly, primarily in the center of the country. This created an extremely crowded area of SAM sites that, for American fliers, constituted a very dangerous MEZ. Inside Iraq there were "roughly" 210 SAMs and 325 combat aircraft, although how many of these were in serviceable condition was unknown. The numbers of aircraft the Iraqi Air Force possessed in 2003 corresponded to what *GWAPS* indicated at the end of Desert Storm with very little discrepancy.[10]

The "Super MEZ" worried coalition air planners and mission pilots alike. Air Forces Central Command chief Lt. Gen. Daniel P. Leaf said, "Countrywide, they are weaker. In Baghdad, they are stronger because they have brought everything in." This included SA-2, SA-3, and SA-6 missiles. At first, the only aircraft initially allowed into the "Super MEZ" were low observable assets accompanied by cruise missiles. Soon, however, in addition to the F-117, the B-2 Spirit joined in combat against the Iraqis.[11]

The Dora Farm Strike

Operation Iraqi Freedom had "first strikes" on two different nights. In the early morning hours of 19 March 2003, two F-117s, flown by Lt. Col. Dave Toomey and Maj. Mark Hoehn, attempted a decapitation strike against Saddam and his sons at the Dora Farm complex. They hit their target but missed the Iraqi leader. Later evidence suggested Saddam had not visited that location in years. The strike was planned and executed in a matter of hours. The previous evening found Toomey and Hoehn eating dinner at the dining facility. Another squadron pilot entered the dining facility and told them to get over to the squadron—something had come up.

On 17 March, the mission planning cell for the F-117s had received a phone call from the CAOC informing them of a potential strike, which was canceled shortly thereafter. The following night, Lieutenant Colonel Toomey was eating a midnight breakfast when he was told to report to the mission planning cell. Arriving at the squadron headquarters, Toomey called Maj. Clint Hinote, the F-117 liaison officer at the CAOC, who told him, "We may have something for you," and sent over target coordinates and photographs. Toomey grabbed Major Hoehn and both of them prepared to step to the jet. "I had no flight plan, no target folder, no nothing."[12]

Luckily, the two squadrons of F-117s, based out of Holloman Air Force Base, New Mexico, had spent the previous several months making the mission planning process for F-117s more flexible and responsive to the possibilities of attacking time-sensitive targets. Previously, it took twenty-four to forty-eight hours to properly plan a strike mission for the stealth aircraft. The streamlined planning proved to be a fortuitous development.[13]

Toomey and Hoehn went to their respective aircraft at around 0245 as the flight crews prepared them for flight, loading two EGBU-27s into each bomb bay. All either pilot knew at this point was that they had a 0530 time over target, which meant departing Al Udeid Air Base no later than 0330. At 0300, another pilot brought them their mission materials: the maps and papers they would need with them in the cockpit to find their target. The mission planning had been rushed, but it was by no means shoddy or slapdash. It was, however, up to the pilots to fly their own routes into Baghdad, which involved finding a refueler in Kuwait before heading into Iraq. At 0305, the pilots started their aircraft and taxied out.[14]

Hoehn recalled, "We were briefed that we would taxi and take off 'comms out,' not talking to anybody. . . . We had folks in the tower that flashed a light gun at us, signaling when we were able to taxi and take off. We went completely stealth, with our antennas in, and anything that would omit a signal turned off."[15]

The two aircraft rocketed into the night and met up with a KC-135 along with support aircraft, including EA-6Bs and F-16CJs. They were cutting their time close, and Toomey told the refueler pilot to turn north and "drag" them into Iraq. The large refueler flew into Iraqi airspace for about twenty miles before the F-117s disconnected and continued north toward Baghdad. Toomey's route took him up the west side of Iraq while Hoehn flew east of the city. They attacked the target from opposite directions. A low cloud cover prevented Toomey from using his laser designator, so he dropped his bombs using GPS. He later recalled, "As I'm driving in, there is a low cloud deck and I could only intermittently see the ground. I thought, 'This is not optimum.'"[16]

By this point, at shortly after 0530, it was light out. Toomey navigated home avoiding large cities where AAA or SAMs might lay in wait. After refueling and landing at Al Udeid, Toomey noticed two security forces troops waiting at the end of the runway for him. He thought to himself, "I probably schwacked the only Catholic church in Iraq." He knew all was well when the two airmen rendered a salute.[17]

If nothing else, the Dora Farm strike proved that American airpower could be enormously flexible and responsive given the right circumstances and that airpower was capable of strategic-level impacts. This strike demonstrated the first use of the EGBU-27 in combat; it had only recently been tested by an F-117 at Holloman. The Dora Farm strike also proved the continued efficacy of stealth technology against the right targets. Low observability continued to provide certain options that no other air force on the planet had and against which very few air defenses had a credible response. Perhaps the greatest contribution of the raid was its demonstration of the importance of flexibility and adaptiveness on the part of the pilots and aircrews. Although it is generally pointless to explore counterfactuals, one wonders what the conflict might have looked like had Saddam been killed that March night. The use of the F-117s in the Dora Farm strike proved the aircraft could be readied for attack in a relatively short amount of time. From that point forward, some F-117s remained on an alert status to be sent against time-sensitive targets. Iraqi Freedom would also see the first retargeting by F-117s in the air.

"Shock and Awe"

Two nights after the Dora Farm raid, on 21 March, the air and ground campaigns began in tandem. The phrase "shock and awe" had been circulating within the US Air Force for quite some time before this point, and on this night it ingrained itself in the American vernacular. It was meant to indicate the rapidity with which airpower could paralyze an enemy's ability to respond and harkened back to a Boydian interpretation of airpower theory. The phrase was first seen in a publication from the National Defense University in 1996. Whether or not "shock and awe" was achieved continues to be debated. After the war, great pains were taken to interview the captured commanders and political leaders of Saddam's regime. Reports from these interviews provide the best evidence for the effectiveness of the early aerial attacks. One commander of a Republican Guard unit commented, "The early air attacks hit only empty headquarters and barracks buildings. It did affect our communication switches which were still based in those buildings. We primarily used schools and hidden command centers in orchards for our headquarters—which were not hit. But the accuracy and lethality of those attacks left an indelible impression on those Iraqi soldiers who either observed them directly or saw the damage afterwards."[18]

The message became clear soon enough to the average Iraqi soldier and, by and large, to the more senior leaders. All of them recognized there was no way the Iraqi military was going to be able to put up an effective resistance against the coalition.

Again it was the F-117s flying in the skies over Baghdad. That night, Maj. Steven "Cruiser" Ankerstar, deputy mission commander for the first wave of strikes, approached Baghdad flying north to south to hit his targets. He saw sporadic AAA and noticed the launch of three SAMs that seemed, to him, to be unguided. Two nights later, on the third night of the war, his cockpit transitioned from dark to bright light as something exploded beneath him. He was never sure what it was, but it did indicate the Iraqis were not going to completely give up.[19]

Twelve years after flying into Iraq for the first time in the cockpit of an F-111, Lt. Col. J. L. Briggs was back in the sky over Iraq, this time in an F-117 on the same mission with Ankerstar. As Briggs recalled, "All the lights were on. I could see cars going down the highway." He described the Iraqi IADS as "fragile." "It wasn't really integrated. The command and control nodes were not effective." By the time Iraqi forces began responding to the explosions occurring in the city, "I was already gone," Briggs said.[20]

The coalition in 2003 launched 700 strikes against more than 1,000 preplanned and targeted aim points called "desired mean point of impact" (DMPI), the actual place on a given target you wanted the weapon to hit. The targets for the opening of the offensive mirrored Desert Storm and Warden's "five rings" methodology for target selection. According to a study conducted by Benjamin Lambeth, these included "suspected leadership locations, regime security, communications nodes, airfields, IADS facilities, suspected WMD sites, and elements of Iraq's fielded forces." Each of these were further broken down into nodes, sites, targets, and eventual aim points. As an example, the communications nodes that first night were broken down into an eventual 112 targets desired to be attacked.²¹

The prohibition of nonstealth assets into the "Super MEZ" did not last long. SEAD attacks quickly rendered the zone inoperable, and air defenses inside this MEZ effectively ended in the first week of the campaign. As victims of both kinetic and nonkinetic strikes, the Iraqis quickly learned to keep their radars off for fear of being targeted. Eventually even slow-flying Predator drones flew through the "Super MEZ" with impunity. Air Forces Central Command kept a meticulous record of enemy responses during Iraqi Freedom: 1,224 reported AAA events, 1,660 rocket and SAM launches, 436 active SAM emitters. Only one coalition aircraft was reported as shot down by either AAA or SAMs.²²

As Lieutenant Colonel Briggs noted, there was a major difference between Desert Storm and Iraqi Freedom: "In 1991, no one had done this before. In 2003, a few people *hadn't* done it. We were as familiar with flying inside Iraq as we were with training ranges back in the States." To some veterans of Northern and Southern Watch or, before that, Desert Storm, Iraqi Freedom was old hat.²³

The Long Wars

The wars in Afghanistan and Iraq continued. Between 2003 and 2011, the US Air Force remained involved fighting two wars simultaneously, neither of which aligned with the USAF's prescribed way of combat. There were no air-to-air engagements and no massive interdiction efforts, in each case, preconceived notions of how to use airpower did not apply to the enemy. It was difficult to apply theoretical concepts—say, Warden's "five rings"—against an enemy whose centers of gravity seemed so ephemeral. I have previously argued in the present work that there no longer existed a true separation of strategic- and tactical-level assets, although clearly both levels of war still existed. The wars in Afghanistan and Iraq saw the use of B-1 bombers in a close air support role, but this fact later became obscured amid a dispute on the home front involving the A-10 and the F-35, with proponents of the former seeming to forget that other assets (e.g., F-16, F-15E, B-1) were capable of performing the CAS mission; moreover, should the USAF become embroiled in a higher-level conflict, the A-10 was not going to be survivable.

After failing to achieve a further status of forces agreement with Iraq, the "last" American jet departed Iraqi airspace in 2011; however, it proved to be only a temporary reprieve before efforts to counter the rise of the Islamic State found American airmen again flying sorties in Iraqi territory. It was airpower against insurgency . . . again.

Combined Joint Task Force–Operation Inherent Resolve (CJTF-OIR)

Nested within CENTCOM, the CJTF-OIR became the lead element in the fight against the Islamic State (referred to variously as IS, ISIS, ISIL, or more derogatorily, Daesh) and conducted strikes in both Iraq and Syria. Despite the already catastrophic civil war occurring in Syria, the United States worked with more than thirty countries in this effort. Beginning in 2014, Operation Inherent Resolve set out to "defeat ISIS in designated areas of Iraq and Syria and set conditions for follow-on operations to increase regional stability." As this book goes to press, Inherent Resolve is ongoing (now in its ninth year), although largely "out of sight and out of mind" of the average American citizen.[24]

Inherent Resolve has been primarily an airpower campaign. In doctrinal terms, the "preponderance of assets" being used in the campaign against ISIS have been the Air Force's. These include the A-10 for CAS missions, as well as the F-16, F-15E, and B-1. These same airframes have also conducted attacks against time-sensitive targets (TST) and preplanned missions against ISIS targets. From the outset, however, the commander of CJTF-OIR has been a US Army officer with an Air Force deputy. The war against ISIS (as of this writing) continues, with airpower and other ground assets slowly gaining and maintaining the upper hand against the enemy. By 2018, no Islamic State strongholds remained in Iraq. However, this "progress" must be overlaid onto a region so war-torn that the term hardly has meaning. The 1991 Gulf War, Northern and Southern Watch, Iraqi Freedom, the rise of the insurgency in Iraq, the Syrian Civil War, intervention by the US-led coalition, a separate intervention by Russia—it can be difficult to tell at any given moment who is fighting whom.

Conclusion

Operations Enduring Freedom and Iraqi Freedom, comprising twenty years of engagements in Afghanistan and nearly the same in Iraq, together with the fight against the Islamic State, demonstrated to the Air Force that its preferred way of combat was certainly not the one most often required of it in an era of lower-level conflict. In such operations as counterinsurgencies, it has often proved difficult for airpower to achieve decisive effects. The USAF certainly demonstrated greater impact in higher-level conflicts. Still, the Air Force proved necessary and in some instances quite decisive in these campaigns. Airpower remained precise and capable of delivering strategic-, operational-, and tactical-level effects. Certainly intelligence and targeting were not always perfect, nor should anyone expect them to be; mistakes and collateral damage are as inevitable in war as they are regrettable. Airpower is an inherently offensive and destructive weapon, regardless of the moral resolve behind its application. It is most effective when used "to kill people and break things," but struggles in nation building and achieving other policy goals. The fall of Afghanistan in 2021 in no way demonstrates a failure of airpower; that outcome is more demonstrative of failures of US foreign policy and long-term strategic planning.

Both Operations Desert Storm and Iraqi Freedom represented the Air Force's preferred way of war. In both instances, the United States faced a modern, industrialized society with a large(ish)—or near-peer—military force; against such, the USAF and its allied airpower participants could leverage technological and training advantages and defeat Iraqi air defense that put up only limited, disorganized opposition, in what resulted in rather lopsided contests in both 1991 and 2003. The ability to plan for and execute an air portion of a campaign using the full weight of America's Air Force—along with allied airpower—remains the cornerstone of the USAF way of war. It is what the USAF continues to train toward through various exercises around the globe, from Red Flag at Nellis Air Force Base to Cope Thunder in the Philippines.

Iraq's procurement of Soviet-built and French-built aircraft and IADS provided the US military the opportunity to ply its trade in a way best understood as the "preferred American way of war." As one general officer put it years later, "It was the major combat operation we always wanted." The Air Force might prefer to be tested in peer-on-peer conflict, but it seems unlikely to get that chance. That has not stopped the US Air Force from attempting to pivot back toward the peer or near-peer threat. Focus on a resurgent Russia (despite its failures in its invasion of Ukraine) and a rising China dominates the thinking inside Air Force planning groups and think tanks. As the future of warfare turns toward proliferation of drones and UAVs, the USAF—and the Department of Defense as a whole—focuses its attention higher . . . toward the final frontier: space.[25]

EPOCH IV

AN UNMANNED FUTURE?
2019–

No matter your thoughts on the subject of unmanned aerial vehicles (UAVs), unmanned aerial systems (UAS), remotely piloted aircraft (RPA), or more commonly if incorrectly drones, it is hard to gainsay they represent a new and interesting role, mission, and platform to be used in aerial warfare, and they have absolutely added a new dimension to its conduct. They look like, as their names imply, a cross between a manned aircraft and a flying computer. While they (typically) possess all the normal aerial attributes: wings, engine, fuselage, and tail, they do not possess a cockpit and that imbues them with a certain undeniably odd aesthetic. Because they do not have a cockpit, they also possess none of the things needed to keep a person alive and safe, such as an environmental control system or an oxygen-generating system; therefore, they also tend to be smaller than their manned counterparts. Those who maintain them have referred to their "china doll" aspect: something that, once broken, cannot be repaired. And while this book argues that, though relatively new to the arena of aerial combat, UAVs are altering our execution of air warfare, "new" is also a problematic term in this context. The long view of unpiloted aircraft has roots in the Second World War. Unmanned craft also demonstrated their ability during the Vietnam conflict, and even in comparison to recent history, say, operations such as Allied Force, vast technological advances are continuing to occur. When the USAF retired the MQ-1 Predator in March 2018, the first true unmanned aerial vehicle to see wide service and combat became a historical artifact.

UAVs are controlled from long distance, from remote locations often on the other side of the globe. This is made possible through a complex system of satellites that are coming under threat by adversaries as space becomes "contested."

188 An Unmanned Future?

As UAVs and space systems proliferate, the very nature of aerial combat is also changing. The USAF is entering a new epoch of airpower, where its future is entirely in question. Thus, this narrative enters its fourth and final epoch, and here the thesis and way of war have yet to be determined. Are UAVs and a newly created United States Space Force truly necessary? Are they fundamentally altering the way of war, or are they merely the latest shiny objects in an inexorable and unalterable way of war? Is the future both unmanned and unbound by the limits of the atmosphere or is it more terrestrial and immutable than some are making it out to be? The pages to follow offer only a brief description of where things stand in 2024, but let there be no doubt: the USAF as an institution certainly perceives its way of war is entering a new epoch.

CHAPTER 9

A FINAL FRONTIER OR AN UNMANNED FUTURE?

The Predator's optics provided a highly magnified image of the tank, but one with a very narrow field of view—similar to what one would see through a soda straw.

—Col. Christopher Haave and Col. Phil Haun, *A-10s over Kosovo*

With NASA growing by leaps and bounds, the Air Force became increasingly edgy about the role it might be assigned in the exploration and exploitation of this new medium grandly called SPACE.

—Michael Collins, *Carrying the Flame*

This fourth epoch and final chapter call into question whether the Air Force's way of war will change with the advent and increasingly widespread use of unmanned aerial vehicles (UAVs), also called unmanned aerial systems (UAS), remotely piloted aircraft (RPA), and, more colloquially and at times erroneously, drones. The role of the United States Air Force operating in and through space also is under the microscope. It is my view that while unmanned aerial systems will make contributions to fighting future conflicts, the military branches of the United States, in particular the US Air Force, remain a long way from a shift to a completely unmanned force, despite the proclivity of some to use the phrase "last manned" for every new fighter or bomber in development. I will also argue that while the Air Force will continue to lead Department of Defense efforts involving operations in space, there will not be a more militarized US presence in that sphere, notwithstanding the establishment of a new United States Space Command (SPACECOM) and the United States Space Force.

Early Development of Unmanned Aircraft with the National Reconnaissance Office and CIA

The rise of armed UAVs altered much of the Air Force's way of war in the last decade. The information age revolutionized what was possible with these systems— namely, that a pilot sitting in an air-conditioned box on one side of the globe could, with the push of a button, send a signal to an unmanned aircraft on the other side of the globe, launch a missile, and bring a life to an inglorious end. Technology, morality, necessity, and psychology are fused into these systems such that we still do not understand and must continue to reckon with. Even from the beginning, there existed a problem with the use of UAVs—namely, that there was no person inside of it. Adoption of UAVs fed into what Thomas Ehrhard called the "organizational schizophrenia of the Air Force." The bottom line was that the Air Force as an institution was initially reluctant to embrace UAVs, but it certainly did not want to leave their development and employment to any other organization.[1]

By whatever name you want to call them, the unmanned aircraft revolution got off to an inauspicious start through the 1960s and 1970s. These weapons systems found themselves pulled between three separate organizations: the Central Intelligence Agency, The US Air Force, and the presumably overarching National Reconnaissance Office (NRO). Programs including Fire Fly/Lighting Bug, Tagboard, Senior Bowl, and Compass Arrow all ended without proving the efficacy of UAVs. Whether because of cost overruns or technological overreach, none of these systems found a lasting presence inside the Air Force. As Ehrhard stated, "The UAV found itself to be a misfit in the increasingly satellite-centered intelligence community, unable to muster consistent support and doomed to a world where the realization of its promise always seemed just out of reach." Still, this did not hamper the desire among some in the Air Force for these systems.[2]

Although contemporary views tend to associate UAVs with twenty-first-century incarnations of airpower, the Vietnam conflict was the first to see extensive use of drone technology. During Vietnam, drones in their embryonic stage of development served as surveillance and reconnaissance assets and also as intelligence-gathering platforms. These UAVs served in part as bait for the SA-2 missile systems, all the while obtaining large amounts of electronic data that the Air Force successfully used to create jamming pods for its manned aircraft.[3]

The Information Revolution Changes UAVs

The massive technological advancements that developed in the 1990s, not least in production capabilities—finally allowed technology to catch up with innovation; as a result, unmanned aerial systems became aircraft capable of providing specific functions—in particular, surveillance and reconnaissance. Thanks in no small part to the Global Positioning System (GPS) and satellite control, which allowed for true over-the-horizon capabilities, UAVs became functioning military systems during the conflicts in the Balkans of the 1990s.

The 1990s Balkans campaigns covered in chapter 7 also saw the first deployment of the RQ-1 Predator. Although the figures are still debated, several Predators were lost during these first deployments, when the aircraft were operated by a combination of military and civilian (CIA) personnel. The higher numbers of AAA faced during the Balkans campaigns resulted in higher loss rates for the UAVs than did later operations in Afghanistan and Iraq. The real proving ground of UAVs from the Air Force's perspective came in the wake of 9/11, so much so that the use of UAVs has become as recognizable as any manned aircraft as being representative of the combat of the last two decades.

Enduring Freedom and Iraqi Freedom

By the time the US military response to the terrorist attacks of 9/11 had begun in October 2001, there was very little doubt that unmanned aircraft would play a major role in the conflict as reconnaissance assets providing near real-time feeds to senior commanders. When the 9/11 Commission Report was issued in 2004, it revealed that Predators were already operating in the sky over Afghanistan more than a year before September 2001, noting that in order to track terrorist cells, "One option was to use a small, unmanned US Air Force drone called the Predator, which could survey the territory below and send back video footage."[4]

Once Operation Enduring Freedom began in October 2001, Predators played an ever-increasing role in the newly dubbed "War on Terror." Just as the UH-1 Huey and the F-4 became emblematic of airpower in the Vietnam War, the Predator proved to be iconic of warfare in the twenty-first century. As the spring of 2002 approached in Afghanistan, the RQ-1 entered a new phase of operations and received a new designation. The now armed MQ-1, capable of firing an AGM-114 Hellfire missile, appeared in the skies over Afghanistan. During one operation as part of Iraqi Freedom, drones played the role of the "Wild Weasels": older model Predators were launched and used as decoys in an effort to tempt Iraqi air defenses to expose themselves by firing.

Inherent Resolve

In the fall of 2011, the United States military was pulled out of Iraq. The author of this work stood on the floor of the Combined Air and Space Operations Center as cheers went up when the last aircraft departed Iraqi airspace. The departure followed the failure of the Barack Obama administration to sign a continuing status of forces agreement with Iraq. It proved to be a short-lived celebration. Again, as noted in chapter 8, the dramatic rise of the Islamic State in Iraq meant that in short order Air Force assets were once again flying in the skies over Iraq. The stand-up of the Combined Joint Task Force–Operation Inherent Resolve (CJTF-OIR) and its ongoing operations again necessitated the employment of both armed and unarmed UAVs.

The future of drones and UAVs is both known and unknown. Proponents note that their operational utility has solidified over the past two decades, from Kosovo to Afghanistan. Opponents and detractors note that UAVs have enjoyed relatively benign environments where they have mostly been allowed to operate with relative impunity, such as in Afghanistan and Iraq, something they did not enjoy over Kosovo and Yugoslavia, but the same could not be said of potential near and near-peer threats. The slow, low-flying nature of the early UAVs indicate they could survive a high-threat environment, but follow-on systems that possess "fighter-like" attributes are expected to fare better. There seems little doubt that UAVs not only will increase in importance, but will likely advance from generation to generation at a faster pace, thanks to the prior development of fighter aircraft along the same lines.

The Final Frontier?

While this work has delved into aerial combat, as well as traditional aerial cargo and other combat support missions, it has, for the most part, eschewed a discussion of the role of space in the Air Force's conception of itself. The USAF has billed itself since at least the early 1960s as America's "air and space" force, in large part because its missile weaponry has been understood as operating *through* space (though not *in* space). Such assets, however, have never been clearly defined as anything other than a support function to the "actual" warfighters. Today, largely because of the general proliferation of satellites in space, and their interconnection to cyberspace, the USAF is now involved in these new potential warfighting domains. As supporters of further military development in the space domain like to point out, the future is not coming, it is already here—and the time to develop further military options in space is now.

Gen. Hoyt Vandenberg, the second chief of staff of the United States Air Force, stated in 1948 that the USAF held the "logical responsibility for the satellite." One might extrapolate a separate space mission from that statement. In fact, in light of recent developments, the USAF now finds a portion of its responsibilities separated into the United States Space Force. This new branch of the armed services has been placed administratively under the Department of the Air Force with its own four-star chief of staff, much the same way the Department of the Navy oversees the United States Marine Corps and the US Army oversaw the USAAF. The influence of space in military operations and the view of space as the final frontier to be explored—some might say conquered—give rise to a discussion of the role of the USAF/USSF as America's space force and an assessment of whether this fits in with the previously discussed ways of war. Does the USAF/USSF space operation fit into a similar way of war, or is it an outlier?[5]

In June 2018, President Donald Trump directed the chairman of the Joint Chiefs of Staff, Gen. Joseph Dunford, to begin work on creating a separate United States Space Force. The order was received somewhat as a bombshell and caught the DOD by surprise. Secretary of the Air Force Heather Wilson sent a letter to the entire Department of the Air Force the next day stating that creating a Space Force was going to be "a thorough, deliberate, and inclusive process. As such, we should not expect any immediate moves or changes." The secretary's statement proved to be wrong. It took only a year for the separate US Space Force to be established as the sixth branch of the United States military, subordinate to the USAF.[6]

On 9 August 2018, Vice President Mike Pence, in a speech at the Department of Defense, announced the establishment (or rather reestablishment) of a functional combatant command for space in advance of the creation of an independent Space Force. President Trump ordered the establishment of SPACECOM in December 2018. The following year saw the issuance of the planning order, and US Strategic Command and the Air Force Space Command (AFSPACE) continued to move toward the establishment of a "functional" space organization inside the DOD at the joint level. All the major pieces were not yet in place by 23 September 2019, when the United States Space Command was reestablished as a combatant command. Less than two months later, an act creating the United States Space Force was signed into law.

The National Defense Authorization Act for 2020 effectively took the Air Force Space Command and "redesignated" it as the United States Space Force (USSF) within the USAF. The leader of the USSF would be designated chief of space operations, and Gen. John Raymond, the former head of Air Force Space Command and SPACECOM, became the first CSO. The function of the new service is to provide "freedom of operation for the United States in, from, and to space; and prompt and sustained space operations." The duties of the service are to "protect the interests of the United States in space and deter aggression in, from, and to space; and conduct space operations."[7]

History of the US Air Force in Space

The United States Air Force space program developed along two parallel paths: (1) unmanned space activities, to include reconnaissance satellites; and (2) a desire for manned space activities separate from those of the National Aeronautics and Space Administration (NASA). This latter desire never came to fruition. However, the imagining of what might be possible played as important a role in concepts of space development as it did in aircraft development.

In much the same way that Jules Verne and H. G. Wells grabbed the imaginations of readers for experiences and adventures in the air in the nineteenth and twentieth centuries, the two authors did the same for the possibility of space travel. In 1865, Verne published *From the Earth to the Moon* (De la terre à la lune), a work that heavily influenced rocket scientists, including Konstantin Tsiolkovsky and Robert H. Goddard. Goddard, Tsiolkovisky, and Wernher von Braun made the idea of rocketry a reality, and thus the possibility that humans might operate in the environment of space.

It was the civilian organization NASA that helped make a reality the already broadly popular ideas of space flight and space exploration. The daring astronauts of the 1960s and 1970s grabbed a powerful hold on the American psyche, much as the World War I aviators had done in 1918. Of the original Mercury 7 astronauts, three were officer-pilots in the USAF: Donald "Deke" Slayton, Leroy "Gordo" Cooper, and Virgil "Gus" Grissom. They were joined in the second group of astronauts by four more Air Force pilots: Frank Borman, James McDivitt, Thomas Stafford, and Ed White. These astronauts were assigned to NASA, but this did not stop the Air Force from attempting to develop its own space program.

Official Air Force operations in space were slow in development. The first definitive steps toward "doing something in space," in the sense that Billy Mitchell described airpower as "doing something in the air," began with the "low-priority, long-term effort" to launch the first American reconnaissance satellite, WS-117L on 11 October 1960. At the behest of the Advanced Research Projects Agency, the Air Force eventually decided to drop the "WS" (which meant "weapons system"). WS-117 moved too slowly for the Eisenhower administration and for the Air Force; it was put on the back burner in favor of projects that could get into space faster.[8]

The USSR got out to something of a head start in the space arena with the deployment of the Sputnik series of satellites. Inadvertently on the Soviets' part, this head start did much to spurn American development in space. As

Secretary of Defense Donald Quarles stated, "The Russians have . . . done us a good turn, unintentionally, in establishing the concept of freedom of international space." In other words, with the Soviet satellites flying peacefully over other countries, there was now an international precedent that a US reconnaissance satellite might fly over the Soviet Union.[9]

In 1959, the USAF moved forward with its own space vehicle by awarding a contract to Boeing Aircraft Corporation for the Dynamic-Soarer, or "Dyna-Soar." In a design that preceded the space shuttle, Dyna-Soar was to be a manned reusable vehicle. The Air Force spent nearly half a billion dollars in development of the spacecraft and in training its own astronauts to fly it, but the Johnson administration killed the Dyna-Soar (designated X-20 in 1962) in December 1963 primarily because the Air Force could not adequately articulate what the mission of the spacecraft was going to be. Initially touting it as a "Hypersonic Glide Rocket Weapon System," the Air Force also believed the X-20 could provide reconnaissance and bombing from space. One of the men chosen in 1960 to train to fly the X-20 was test pilot Neil Armstrong.[10]

In the development of an Air Force space program, 1960 proved to be a pivotal year. In May, Francis Gary Powers was shot down in his U-2 over the Soviet Union, which spurred the desire for reconnaissance satellites that were capable of not being brought down by Soviet SAMs. This was accomplished with the launch of the Discoverer/Corona satellite and the first MIDAS satellite to be used for early warning. Launched between 1959 and 1960, Discoverer 1–12 all failed, but the follow-on missions 13 and 14 finally worked. The latter proved to be something of a harbinger of things to come as the Air Force began using its satellites for missile warning. These operations fell under the previously mentioned mission of NORAD.

Shortly before the cancellation of the Dyna-Soar, the Air Force moved forward on another space project called "Blue Gemini." The program was designed to use spacecraft that were part of the Gemini program with Air Force crews to build a Manned Orbital Development System (MODS), a sort of early space station for military use. This was to be added on top of the already planned twelve Gemini missions. A NASA history stated that "Blue Gemini was neither clearly defined nor officially sanctioned," and that "Air Force opinion was divided on the best approach to the goal of military manned space flight." Of particular importance, the Air Force chief of staff, General LeMay, saw no future in a program that intertwined the USAF with NASA.[11]

Aspects of the Blue Gemini program were met with ire by NASA officials, particularly when Secretary of Defense McNamara proposed moving the entire Gemini program under the DOD. NASA was careful to guard against what it rightly perceived of as encroachment by the Defense Department.

Perhaps the best hope for a manned Air Force program in space came with the Manned Orbiting Laboratory (MOL). Based off use of Gemini capsules attached to a separate laboratory vehicle (a hollowed-out Titan rocket), the MOL called for two astronauts to operate continuously in space for up to thirty days at a time, a considerable feat considering that even later Apollo missions did not approach that duration. The question, however, remained: What experiments would the military officers perform? Clearly, there was the hope of spying on the Soviets, but did that require a manned presence in space? One Air Force history noted that "some in the Department of Defense argued that man had no military role in space."[12]

It was no surprise, then, when all of these programs—Blue Gemini, the X-20, and the Manned Orbiting Laboratory—eventually met their demise, effectively ending any bid by a military service to take a more active role in manned space flight. Most astronauts continued to be drawn from the ranks of the military across all of the services, but these program cancellations ended any movement toward a manned military presence in space. The test pilots chosen to fly on the MOL program contacted NASA after its cancellation in 1969, a mere month before Apollo 11 landed on the moon, and many of these, including Bob Cripen, eventually flew on space shuttle missions. In the coming decades, there were several exclusively military space shuttle missions and some that contained classified cargo deployments, but these were but a handful of the overall 135 shuttle missions. This forced a major change in the direction the Air Force took in relation to future space operations: working through space rather than working in space. This continued for much of the following five decades.[13]

ICBMs

Closely tied to the Air Force's desire for operations in the space domain was the development of intercontinental ballistic missiles (ICBMs). First conceived of during the Eisenhower administration and brought to reality and employment under the leadership of Gen. Bernard Schriever, ICBMs became one leg of the United States' nuclear triad. Thus, the Air Force owned and operated two legs of the triad (ICBMs and manned SAC bombers), while the US Navy's ICBM submarine force constituted the third. The first Air Force missiles—the Atlas and the Titan—were also used by NASA to send early astronauts into space. The Atlas and Titan gave way to the Minuteman and Peacekeeper variants. The SALT II treaty in 1979 led to the demise of the Peacekeeper, and as of 2019, the Minuteman III force remains the sole nuclear deterrent ICBM in use by the USAF. The missile force was a part of the USAF Strategic Air Command until it was inactivated in 1992 with the creation of Air Combat Command.

Less than a year later, the 20th Air Force (effectively the USAF's nuclear missile force) was moved to Air Force Space Command, where it remained until the stand-up and reestablishment of Global Strike Command. The missile combat crews, more often called "missileers," of the USAF have remained "on alert" for the better part of sixty years, but are largely absent from these pages. These men and women created their own unique culture, which has been overshadowed by the "big blue" Air Force, shaped by the difficulty of being an organization whose primary mission is to ensure the avoidance of using its primary weapon.[14]

However, despite their accomplishments and the USAF's own missile programs, the idea of an independent working space for space officers did not exist until 1 September 1982 with the creation of AFSPACE. The proliferation of individual objects in space and the expansion of satellite constellations throughout the 1960s and 1970s necessitated an organizational change. In order to manage the expanding field of military operations through and in space, the United States Space Command was established in 1985. The commander in chief of US Space Command (CINCSPACE) was "triple-hatted" in that he served not only as CINCSPACE, but also as commander of AFSPACE and as CINCNORAD as well. This arrangement lasted until 2002 when US Space Command was disestablished. Certain functions of the command were split between Strategic Command and AFSPACE. The binational command of NORAD was then attached to the newly created United States Northern Command. This relationship remained in effect even as the Trump administration moved forward with the stated goal of standing up the United States Space Force by 2020. It remains to be seen whether Congress will provide adequate funding for the new military branch.

Expansion of the Air Force in Space

One of the greatest technological changes to warfare occurred with the advent of the Global Positioning System. Beyond applications for location and navigation, GPS eventually found its way into munitions, to the point in the 2000s that unguided weapons became all but obsolete. As of 2018, utilizing thirty-one operational satellites—twenty-four must be operational at any given moment—located in medium earth orbit (MEO), the constellation allows for munitions leaving the rails to—for the most part—hit within feet of their aim point. The USAF currently maintains and monitors around three dozen satellites. The access to GPS has become virtually ubiquitous, such that anyone with a cell phone expects to be able to pinpoint their exact location at any given moment (despite the fact that this access normally comes from triangulated cell phone towers, it is still referred to as "GPS service").[15]

Beyond the actual space assets, there is a "control segment." The master control station (MCS) is located at Schriever Air Force Base outside of Colorado Springs, Colorado. The MCS oversees the nearly thirty ground antennas, Air Force monitor stations, Air Force Satellite Control Network remote tracking stations, National Geospatial-Intelligence Agency monitor stations, and one alternate master control station. Two Air Force squadrons monitor and maintain the GPS constellation to, in their own words, "keep the GPS satellites flying."[16]

What distinguished the idea of a separate "Space Force" from the similar move for an independent Air Force was the complete lack of senior leaders or lower- and midlevel officers agitating for independence. Although it was widely recognized that the "US military gradually became dependent on space assets for operational wartime support," it was equally recognized that the space mission was simply that—one of support. Officers in the Air Force's Space Command were derisively called "penguins" by pilots, meaning that they had wings but could not actually fly.[17] To counter this perception, the Air Force looked to "operationalize, institutionalize, and normalize" activities of the Air Force Space Command, or to make AFSPACE airmen look like its pilots in other major commands.[18]

Conclusion

Although some of this chapter has examined the past history of the Air Force in space, it is risky for a historian to make pronouncements about the future, not only of the United States Space Force itself but with respect to how the USAF will handle the establishment of the new service. The mission of the USSF remains vaguely defined and not entirely understood by the American people it serves. The bottleneck, however, continues to be lack of defined missions outside of vague pronouncements of about countering what possible adversaries are capable of doing—or what they might be capable of doing in the future—in the domain of space. Operating in space remains a hugely expensive venture, even as more public companies, such as SpaceX and Blue Origin, have begun operating there.

In the spring of 2017, Congressman Mike Rogers of Alabama floated the idea of establishing a USAF Space Corps by 1 January 2019, with a Space Corps chief of staff reporting directly to the secretary of the Air Force as a coequal with the chief of staff of the Air Force and appointed for a term of six years as a member of the Joint Chiefs of Staff. The use of the word "coequal" seemed to be a deliberate nod to FM 100–20, which established as official Army doctrine that "land power and air power are co-equal and interdependent."[19] This seemed to be the direction a possible space force was taking as well.

The Air Force immediately balked, with Secretary of the Air Force Dr. Heather Wilson saying, "The Pentagon is complicated enough. We're trying to simplify. This will make it more complex, add more boxes to the organization chart and cost more money. If I had more money, I would put it into lethality not bureaucracy." Secretary Wilson's statements and clashes with the Trump administration, in particular with the president himself, led to her resignation in May 2019.[20]

The initial idea of a separate Space Corps did not survive the 2017 National Defense Authorization Act. However, the questions raised about the future militarization of space clearly drew new lines in the sand regarding the very idea of a way of war for the USAF. It did not take long for the Trump administration to cross these lines with the intention of standing up a separate combatant command for space by the end of 2018. As discussed earlier, all of this came to fruition in 2019. It seems the future of the United States Space Force—how it will be manned, how large the organization will be, what its uniforms will look like—are all questions yet to be fully answered.[21]

Conclusion

The history of the USAF told herein is that seen from the 50,000-foot level—a broad and wide-ranging view, but one capable of seeing the details as each aircraft, individual persona, event, or conflict is passed over. In a book review published in the *Journal of Military History* in January 2019, Sebastian Lukasik stated, "Introductory surveys of airpower history are plentiful. To stand out in this crowded field, a book must offer something more than the standard chronological march from the Wright Flyer (or the Fleurus balloon, if one is tempted to go back further into the past) to the Predator drone, and conceptualize airpower in terms that go beyond the traditional focus on the advancement of technological capabilities as the driving force behind developments in air warfare."[1]

This review, published just as this volume was entering the editing process, proved an apt warning for the author. What had this present work done significantly different, in light of Lukasik's critique?

This book looks at the Air Force's way of war from the first military Wright Flyer of 1909 to the specialized unmanned aerial vehicles in the second decade of the twenty-first century, exactly the period Lukasik warns against. It covers engagements, conflicts, and wars from the Army Signal Corp's earliest expedition against Pancho Villa to the recent Operation Inherent Resolve. Those two operations had more in common than some might like to admit. Both involved an enemy operating in a familiar environment and using its asymmetric advantages to stymie the efforts of the American military. In both cases, airpower provided intelligence, reconnaissance, and surveillance, but in the latter operation, the platforms were unmanned and many carried a retaliatory capability of their own. Ironically, though they bookend this history, neither represents the way the Air Force views itself.

This book demonstrates four definitive ways of war for the USAF: a period of discovery including development and growth, eras of strategic dominance and tactical ascendancy, and a possibly unmanned or space-based operational paradigm, which may come to define a new way of war. This book also demonstrates that a unifying feature in every epoch is and always has been the people, the fliers and mechanics, of America's air arm. While platforms, technology, roles, missions, and systems have changed, it has always been the men and women willing to climb inside these machines and take the fight forward who truly represent the Air Force's way of war.

The Air Force has always sought "victory through airpower," and it has always sought solutions to problems by going "over and not through." This was true in the earliest Nieuports, SPADs, and Sopwith Camels and is still true in the more modern incarnations of airpower. The desire of those who sought to lessen the carnage of war drove them to devise a strategic bombing theory that, perversely, only increased the horrors of warfare, but that same drive and determination led to the creation of precision-guided munitions. No one has ever rightly claimed that aerial warfare would be cleaner or less horrendous than the conflicts that came before the advent of the airplane, but the drive to end wars sooner and to bend technology to human purposes when it comes to war has played an overarching role in how the airplane developed.

Along the way, altitude gave way to orbit as the organization, not always successfully, moved higher, farther, and faster. From biplanes to bombers; from Sopwiths to stealth fighters; the Air Force proactively sought to utilize airpower in ways that could, while contributing to the fight, shorten or even decisively end a conflict. Not everything promised came to pass, and not everything that came to pass was envisioned before. Technological determinism and technological innovation both have a place in the history of the US Air Force.

Technology and Platforms

The at times vicious debates about the effectiveness of airpower, which have accompanied every engagement it has ever participated in continue in the present. From a certain point of view, the current debates reprise those that came before, and they all, in some way, become binary "either/or" arguments in which context and nuance are lost. In 2017, it was the F-35 versus the A-10. Although the platforms performed vastly different missions, reductive arguments made the conversation about which aircraft did one *single* thing (that is, close air support) better than the other. In the early 2000s, it was the utility of the F-22. In the early 1980s, it was the problems plaguing the development of the F-16. In between these arguments came shining examples of airpower dominance and success: the strikes against Libya in 1986, Desert Storm in 1991, followed by the Balkans campaigns and the melding of Special Forces troops on the ground with dominant airpower as in the early stages of Enduring Freedom.

There were failures too. Preconceived notions of how to apply airpower met stiffer than expected resistance during the Vietnam conflict. Technological advance received a spear to the side when an F-117 was shot down in 1999.

There were also failures in theory. The idea—somewhat misattributed and misunderstood— that the bomber will always get through was proven not entirely correct over the skies of Schweinfurt and Ploesti when aircraft and aircrew losses mounted. Saying "not entirely correct" is not meant to obfuscate, as airmen generally believed that a bombing mission, once begun, could not be stopped. Even naval aviators were guilty of such beliefs, including the father of carrier airpower Adm. William Moffett, who stated, "I am convinced that a bombing attack launched from such carriers [the *Lexington* and *Saratoga*] from an unknown point, at an unknown instant, with an unknown objective, cannot be warded off."[2]

The Innovators

The Air Force has never truly rewarded innovative or "outside the box" thinking. Many of those held in high esteem today, including Billy Mitchell and Claire Chennault, were in their day fired or chastised for their innovative thinking that went against the established cultural norms. Col. John Warden, after masterminding the structural framework for the air campaign in Operation Desert Storm, was moved to be commandant of the Air Command and Staff College, and he never made general despite his personal impact on how the Air Force conceived of and executed aerial warfare in the 1990s and early 2000s. To go against institutional norms is to question the institution as a whole. As Donald Mrozek once said, "Yet a key to the innovating and venturesome spirit is the willingness to rewrite the rules. But the 'nonconventional person who's going to do it differently' becomes 'a flat-out pain in the ass' for the guardians of institutional order, hierarchy and deference, and the status quo." For every good decision made by the Air Force, someone had to be sacrificed on the altar of old-fashioned order and tradition and martyred to prevent upsetting the institution.[3]

That being said, Air Force is replete with lesser-recognized innovators who had an impact on the conduct of aerial war. The members of the Lafayette Escadrille of World War I were willing to leave country behind in service to something greater than themselves. They then trained and mentored a nascent American Air Service in 1918. George Kenney, in the Southwest Pacific during World War II, used whatever means at his disposal to prosecute his corner of the conflict. William Tunner brought forth an entire generation of air transport professionals. The WASPs took the first tentative steps toward women's service, and names like Nancy Love and Jackie Cochran went down in history, making possible the rise of Gen. Lori Robinson, the first female to lead a combatant command, from 2016 to 2018, and others, whose impact surely inspired another generation. The Tuskegee Airmen fought for victory overseas and then for full civil rights for all at home. The generation after Vietnam, including Chuck Horner, Moody Suter, and Robbie Risner, made concrete changes to training and operational paradigms that are still felt by the USAF today.

Every martyr had acolytes, and those acolytes were willing to "take up the cross" of their leader and bring his dreams to fruition, as the case of Billy Mitchell so dramatically indicated. Once the dream was fulfilled, the old followers, now the leaders of the institution, guarded their power against the inevitable next wave who would want to change the old way of doing things that had so recently seemed new and even revolutionary. So strong were the influences of the "bomber mafia" that they held sway for decades. As Stephen Harris and Robert Higham commented, "Such was the philosophical essence of Giulio Douhet, Hugh Trenchard, and Billy Mitchell, the main theorists emerging from the First World War—an essence so powerful that nations that could not afford to build a bomber force (or had no targets for one) nevertheless became entranced by it, often at the expense of other elements of what we would now call aerial warfare."[4]

The United States Air Force and its predecessor organizations have conducted aerial warfare for more than one hundred years. The centenary of American airmen conducting operations in the First World War occurred in 2018. Some of the pursuit squadrons that participated at Saint-Mihiel and in the Meuse-Argonne campaign became today' fighter squadrons, having traded their Nieuports and SPADs for F-22s. Speed, range, payload, armament—all are different, but some of the tactics used and maneuvers performed retain the names their inventors gave them.

In the end, the question remains: what is the way of war for the United States Air Force? It is, at once, ever-changing and never-changing. America's Air Force saw three distinct epochs of airpower—early development, strategic dominance, tactical ascendancy—and has entered its fourth: the era of unmanned aerial vehicles and a future in space. The platforms of the present time are wholesale different from those in the early twentieth century, but the drive of airmen to push the bounds of what is technologically feasible is the same. The way of war is individuals, serving in teams, using history as a guide, but always looking forward. "Over not through" continues to move the organization toward its ultimate goal of "victory through airpower." Even with future moves toward as-yet undesigned next-generation aircraft, one thing remains clear: the United States Air Force will do what it has always done: in the memorable words of Billy Mitchell, the USAF will continue to "do something in the air."[5]

Note on Sources and Acknowledgments

This book, a general and historical synthesis of the United States Air Force, is based almost entirely on the work of other historians, fliers, and scholars. Without their diligence and heavy lifting, this book would never have gotten off the ground. I have attempted to point toward their work throughout, at the very least in the notes, if not directly in the text itself. I am heavily indebted to the work of the airmen and scholars who came before me whose work made this book possible. The conclusions and historical analysis are mine alone.

The Air Force History and Museums Program (http://www.afhistoryandmuseums.af.mil/) is charged with the mission of preserving USAF history. As such, many of the original sources and secondary books came from this program and their subordinate offices including:

Air Force Historical Research Agency (AFHRA) (http://www.afhra.af.mil/)
Air Force Historical Support Division
United States Air Force Major Command History Offices
    Air Combat Command Office of History
    Air Mobility Command Office of History
    Air Force Space Command Office of History
    Global Strike Command Office of History
    Air Force Special Operations Command Office of History
United States Air Force Academy, Clark Special Collections Branch

Other historical collections and archives used include:

University of Virginia Library, James Rogers McConnell Memorial Collections
Rutgers University, School of Arts and Sciences, Oral History Archives, Frederick Wesche III interview
NASA Office of History

Still, there are individuals who merit special attention: Mary Dysart, Daniel Haulman, and Samuel Shearin at the AFHRA. I need only send an email and the document requested magically showed up in my inbox. Adam Kane, my initial editor at the University of Oklahoma Press, who thought of me and initially approached me about doing this book: Thank you, I think. The Ways of War series editors, David J. Ulbrich and Matthew S. Muehlbauer. The editorial and production staff at the University of Oklahoma Press, including Andrew Berzanskis, Tawna Dickens, Amy Hernandez, and Helen Robertson. Others at OUP have moved on, but they deserve my thanks as well: James K. Calder and Katie Hall. Special thanks to my freelance copy editor Robert Fullilove and my indexer, Galen Schroeder. Daniel Wheaton and Gerry White, both excellent Air Force historians at Nellis Air Force Base; Paul Springer and John Terino at the Air Command and Staff College. My fellow editors and friends at Balloons to Drones: Ashleigh Brown, Maria E. Burczynska, Alex Fitz-Black, Michael Hankins, Ross Mahoney, Victoria Taylor, and Luke Truxal. Special thanks go to Dr. Hankins, who also served as a peer reviewer for this work. His extensive suggestions and recommendations are the only reason this book ever saw the light of day.

Several veterans of the conflicts covered in these pages helped with personal stories, including J. L. Briggs and Dave Toomey, who consented to a series of interviews about their experiences flying the F-117. Finally, I must acknowledge the many thousands of #twitterstorians I have never met but interact with on a daily basis. If not for you . . . this book would have been done a lot sooner.

Notes

Introduction

1. Colin S. Gray, *Airpower for Strategic Effect* (Maxwell AFB, AL: Air University Press, 2012), 272–75.
2. Antulio J. Echevarria II, *Reconsidering the American Way of War: US Military Practice from the Revolution to Afghanistan* (Washington, DC: George Washington University Press, 2014), 162; Thomas S. Kuhn, *The Structure of Scientific Revolutions*, 4th ed. (Chicago: University of Chicago Press, 2012).
3. Alexander P. de Seversky, *Victory through Air Power* (New York: Simon and Schuster, 1942).
4. For some of the "ways of war" scholarship, see Russell F. Weigley, *The American Way of War* (Bloomington: Indiana University Press, 1973), xxii; Echevarria, *Reconsidering*; Robert M. Citino, *The German Way of War: From the Thirty Years' War to the Third Reich* (Lawrence: University Press of Kansas, 2005); John Grenier, *The First Way of War: American War Making on the Frontier, 1607–1814* (Cambridge: Cambridge University Press, 2005); Patrick M. Malone, *The Skulking Way of War: Technology and Tactics among the New England Indians* (Lanham, MD: Madison Books, 1991); and Brian D. Laslie, *The Air Force Way of War: U.S. Tactics and Training after Vietnam* (Lexington: University Press of Kentucky, 2015). For airpower and USAF history, see Frank Ledwidge, *Aerial Warfare* (Oxford: Oxford University Press, 2018); Charles Gross, *American Military Aviation* (College Station: Texas A&M University Press, 2002); John Andreas Olsen, ed., *A History of Air Warfare* (Washington, DC: Potomac Books, 2010); Matthew S. Muehlbauer and David J. Ulbrich, *Ways of War*, 2nd ed. (Washington, DC: Potomac Books, 2018).
5. Gen. James P. McCarthy and Col. Drue L. DeBerry, eds., *The Air Force* (Clinton, MD: Air Force Historical Foundation and Hugh Lauter Levin Associates, 2002); Bernard Nalty, ed., *Winged Shield, Winged Sword*, 2 vols. (Washington, DC: Air Force History and Museums Program, 1997); Frank Futrell, *Ideas, Concepts, Doctrine: Basic Thinking in the United States Air Force*, 2 vols. (Maxwell AFB, AL: Air University Press, 1989).
6. Paula G. Thornhill, "'Over Not Through': The Search for a Strong, Unified Culture for America's Airmen" (Santa Monica, CA: RAND Corporation, 2012); de Seversky, *Victory through Air Power*.

7. Organizational Records Division, https://www.afhra.af.mil/Information/Air-Force-Organizational-Records/, Air Force Historical Research Agency (hereafter cited as AFHRA); Georges Thenault, *The Story of the Lafayette Escadrille* (Boston: Small, Maynard & Company, 1921), 3–4.
8. William Mitchell, *Winged Defense* (1925; reprint, Mineola, NY: Dover, 2006), xii; James J. Cooke, *Billy Mitchell* (Boulder, CO: Lynne Rienner, 2002), xii; Alfred H. Hurley, *Billy Mitchell: Crusader for Air Power* (Bloomington: Indiana University Press, 2006); Roger G. Miller, *Billy Mitchell: "Stormy Petrel of the Air"* (Washington, DC: AFHMP, 2004). For a work demonstrating airpower's unexplored limits, see Mark Clodfelter, *The Limits of Air Power* (New York: Free Press, 1989). For examples of unvarnished praise for airpower, see Richard Hallion, *Storm over Iraq* (Washington, DC: Smithsonian Institution Press, 1992); Cooke, *Billy Mitchell*; Hurley, *Billy Mitchell*; Miller, *Billy Mitchell*.
9. Mitchell, *Winged Defense*, vii–viii.
10. Mauer Mauer, ed., *The U.S. Air Service in World War I* (Washington, DC: Office of Air Force History, 1979), 4 vols., 4:67.
11. Lee Kennett, *The First Air War 1914–1918* (New York: Free Press, 1991), 217; Jean-Marc Marill, *Aéronautique*, quoted in Kennett, 218.
12. Mitchell, *Winged Defense*, xvi.
13. The Air Corps Tactical School was also known in several different iterations as the Air Service Field Officers School and Air Service Tactical School. For simplicity's sake, I have used Air Corps Tactical School throughout. Alan Stephens, ed., *The War in the Air 1914–1994* (Maxwell AFB, AL: Air University Press, 2001), 38; Maurer Maurer, *Aviation in the U.S. Army 1919–1939* (Washington, DC: Office of Air Force History, 1987), 65. A note on Carl Spaatz: in 1937, at the urging of his wife and daughters, he added an extra *a* to ensure it was pronounced as in "spa" not "spat." For clarity and simplicity's sake, I have chosen to use Spaatz throughout. Kuter commented on Chennault's ouster from ACTS: "We just whipped him." Gen. Laurence Kuter, oral history interview [hereafter cited as OHI], vol. 1, 111, AFHRA.
14. Stephens, *War in the Air*, 38; Keith Middlemas and John Barnes, *Baldwin: A Biography* (New York: Macmillan, 1979), 735. At the time, Lord Stanley Brown was lord president of the council.
15. The bombardment of industrial targets was designed to inflict damage on the enemy's ability to wage war. The concept was introduced at the Air Corps Tactical School as part of the "industrial web theory" and bears many similarities to John Warden's "five-ring concept" and "effects-based operations" discussed later in this book; Giulio Douhet, *The Command of the Air* (Washington, DC: Office of Air Force History, 1983), 58.
16. Martha Byrd, *Chennault* (Tuscaloosa: University of Alabama Press, 1987), 40.
17. Air Corps Tactical School, Employment of Combined Air Force, 1926; Robert T. Finney, *History of the Air Corps Tactical School, 1920–1940* (Washington, DC: Air Force History and Museums Program, 1998 [1955]); Byrd, *Chennault*, 40, 61; Thomas A. Hughes, *Over Lord* (New York: Free Press: 1995), 57.
18. De Seversky, *Victory through Air Power*, 254. The United States Army Air Corps had been reestablished as the United States Army Air Forces in June 1941.
19. David Brown, "Victory through Hot Air Power," *"PIC,"* 15 January 1943, 7.

20. Edward Warner, "Douhet, Mitchell, and Seversky: Theories of Air Warfare," in Edward W. Earle, ed., *Makers of Modern Strategy: Military Thought from Machiavelli to Hitler* (New York: Athenaeum, 1969), 502.
21. Russell E. Lee, "Impact of Victory Through Air Power," *Air Power History*, 14;. The beard quote also was also mentioned in Brian D. Laslie, *Architect of Air Power* (Lexington: University Press of Kentucky, 2017), 95–96.
22. Finney, *Air Corps Tactical School*, 33; 56.
23. Laslie, *Architect of Air Power*, 39.
24. Barton C. Hacker, *American Military Technology: The Story of a Technology* (Baltimore: Johns Hopkins University Press, 2006), 101.
25. Arnold, quoted in Phillips Payson O'Brien, *How the War Was Won*, 456.
26. *The United States Strategic Bombing Surveys* (Maxwell AFB, AL: Air University Press, 1987 [1945]) [hereafter cited as *USSBS*], 18–19; Hughes, *Over Lord*, 13. Bomber aircraft dropped more than 1.4 million tons of bombs in Europe alone, but the surveys do not note how much tonnage tactical aircraft dropped.
27. *USSBS*, 82–86.
28. *USSBS*, 106.
29. Thomas E. Griffith Jr., *MacArthur's Airman* (Lawrence: University of Kansas Press, 1998), 15; George C. Kenney, *General Kenney Reports* (Washington, DC: Office of Air Force History, 1987), 569. Examples of the focus on strategic bombardment include: Geoffrey Perrett, *Winged Victory* (New York: Random House, 1993); Kenneth Werrell, *Blankets of Fire: U.S. Bombers over Japan during World War II* (Washington, DC: Smithsonian Institution Press, 1996); Barrett Tillman, *Whirlwind: The Air War Against Japan 1942–1945* (New York: Simon and Schuster, 2010).
30. Thomas E. Griffith, *MacArthur's Airman* (Lawrence: University of Kansas Press, 1998), 15.
31. *USSBS*, 37.
32. Hughes, *Over Lord*, 13; Michael S. Sherry, *The Rise of American Air Power* (New Haven, CT: Yale University Press, 1987), ix.
33. Air Combat Command Office of History, Strategic Air Command File Box. The US Air Force breaks its subcomponents into "major commands," commonly referred to as MAJCOMs. The major commands are grouped by mission type so that, at the creation of the Air Force, the strategic bombers were all in Strategic Air Command and the tactical fighters in Tactical Air Command. The function of airlift was handled my Military Airlift Command. The commands have changed many times since 1947 but nomenclature of each command has always given a good indication of what type of aircraft and personnel are typically found in each command. Although each MAJCOM reports to the headquarters staff of the US Air Force they work in a quasi-autonomous state.
34. National Security Council document NSC 162-2, https://irp.fas.org/offdocs/nsc-hst/nsc-162-2.pdf; Conrad C. Crane, *American Air Power Strategy in Korea 1950–1953* (Lawrence: University of Kansas Press, 2000), 21. For more commentary see Trevor Albertson, *Winning Armageddon: Curtis LeMay and Strategic Air Command, 1948–1957* (Annapolis: Naval Institute Press, 2019); and Melvin G. Deaile, *Always at War* (Annapolis: Naval Institute Press, 2018).
35. Crane, *American Air Power Strategy*, 172.

36. Crane, *American Airpower Strategy*, 49.
37. Air force leaders were not alone in this respect. The entire Department of Defense focused Europe at the expense of the actual combat occurring in Korea at the time. For LeMay's comments see Crane, *American Airpower Strategy*, 26.
38. The F-86 was the last aircraft designed primarily for air superiority until the debut of the F-15 in the 1970s. The US Air Force pilots faced North Koreans, Chinese and Soviet pilots over MiG Alley.
39. Crane, *American Air Power Strategy*, 175. It is difficult to accurately state kill-ratios for different conflicts. One must take into account that there are different allies and different services, participating in each conflict and each of these has different qualifications necessary to earn a "kill." In Korea the United States Air Force claims to have shot down 898 enemy aircraft of all types while only losing 78 "American aircraft," and does not differentiate between types of aircraft (fighter, reconnaissance, bombers, etc.) This is roughly a 9–1 ratio. The Soviets, North Koreans and Chinese claimed to have shot down several hundred American aircraft. The huge disparity in numbers is disconcerting but the 7 to 1 ratio quoted above is as accurate as is reasonably possible. Source: http://afhra.maxwell.af.mil/avc_query.asp?SUM= War and http://www.afhra.af.mil/aerialvictorycredits/index.asp.
40. Crane, *American Air Power Strategy*, 50.
41. Kennett, *First Air War*, 219; Kenneth Werrell, "The Air Force and the Future of the Strategic Bomber," *Air University Review* 27, no. 6 (1976): 73–80.
42. Donald J. Mrozek, "In Search of the Unicorn: Military Innovation and the American Temperament," *Air University Review* 37, no. 6 (1986): 28–45.
43. It is recognized that other CSAF's had fighter experience including the first CSAF Carl Spaatz, but Gabriel is generally accepted as the first CSAF to be a part of the new "Fighter Mafia."

Epoch I

1. Graham Rood, "A Brief History of Flying Clothing," *Journal of Aeronautical History*, Paper No. 2014/01, 7.

Chapter 1

Squier, quoted in James J. Cooke, *The U.S. Air Service in the Great War, 1917–1919* (Westport, CT: Praeger, 1996), 8.

1. Edwin W. Morse, *America in the War* (Charles Scribner's Sons, 1919), https://net.lib.byu.edu/estu/wwi/comment/AmerVolunteers/MorseTC.htm, Part VI, Chap. XXVII.
2. Martin P. Claussen, *Materiel Research and Development in the Army Air Arm: 1914–1945*, Army Air Force Historical Studies no. 50, AAF Historical Office, November 1946, 11, AFHSO.
3. Juliette Hennessy, *United States Army Air Arm, April 1861–April 1917* (Washington, DC: Office of Air Force History, 1985), 29, 33.
4. "Diary of V. L. Burge 1909–1912," and Col. V. L. Burge, "Early History of Army Aviation," 7, AFHSO.
5. Kuter, OHI, 1:299; Benjamin D. Foulois, *From the Wright Brothers to the Astronauts* (New York: McGraw-Hill, 1968), 68.
6. Hennessy, *Army Air Arm*, 168.

7. Alejandro de Quesada, *The Hunt for Pancho Villa* (Oxford, UK: Osprey Publishing, 2012), 33.
8. David R. Mets, *Master of Airpower* (Novato, CA: Presidio Press, 1997), 17,19. At the time of the Punitive Expedition, Spaatz still spelled his name with the single *a* as Spatz. It was not until many years later at Langley Field in Virginia around 1937 that the second *a* was added. Mets, 104, 344.
9. Foulois, *From the Wright Brothers*, 127, 128, 131.
10. Foulois, *From the Wright Brothers*, 127, 128.
11. Foulois, *From the Wright Brothers*, 136; Hennessy, *Army Air Arm*, 172.
12. Johannes Werner, *Boelcke der Mensch, der Flieger, der Führer der deutschen Jagdfliegeriei* [*Knight of Germany: Oswald Boelcke—German Ace*] (Leipzig: K. F. Koehler Verlag/New York: Arno Press, 1972), 183–84; R. G. Head, *Oswald Boelcke* (London: Grub Street Publishing, 2016), 97–98.
13. Olsen, *History of Air Warfare*, 12.
14. Kennett, *First Air War*, 73. Boelcke died after an aerial collision with another German fighter in 1916 at the age of twenty-five; McCudden died two years later in July 1918 at the age of twenty-three.
15. Craig F. Morris, *The Origins of American Strategic Bombing Theory* (Annapolis: Naval Institute Press, 2017), 6.
16. H. G. Wells, *The War in the Air* (New York: Grosset & Dunlap, 1908). Jules Verne's *From the Earth to the Moon* would equally influence early rocket theorists, just as Wells influenced airpower theorists.
17. Tami D. Biddle, *Rhetoric and Reality in Air Warfare* (Princeton, NJ: Princeton University Press, 2002), 38–39.
18. Hennessy, *Army Air Arm*, 197.
19. Claussen, *Materiel Research and Development*, 12.
20. Morse, *America in the War*, part VI, chap. XXVII.
21. Thenault, *Lafayette Escadrille*, 2–3, 9; Morse, *America in the War*, part VI, chap. XXVII.
22. Morse, *America in the War*, part VI, chap. XXVII; Thenault, *Lafayette Escadrille*, 15.
23. Numerous bureaucracies inside the French government had to be navigated in the creation of an American flying squadron under control of French officers, and it never would have come into being without the help of men such as M. de Sillan and Dr. Gros, the latter the head of the Automobile Ambulance Section. Morse, *America in the War*, part VI, chap. XXVII; Thenault, *Lafayette Escadrille*, 14.
24. Thenault, *Lafayette Escadrille*, 8.
25. Thenault, *Lafayette Escadrille*, 4, 9.
26. Roger G. Miller, *Like a Thunderbolt: The Lafayette Escadrille and the Advent of American Pursuit in World War I* (Washington, DC: Air Force History and Museums Program, 2007), 1.
27. Thenault, *Lafayette Escadrille*, 45.
28. Thenault, *Lafayette Escadrille*, 50.
29. Thenault, *Lafayette Escadrille*, 56, 66–67.
30. Thenault, *Lafayette Escadrille*, 65; Bert Frandsen, *Hat in the Ring* (Washington, DC: Smithsonian Institution Press, 2003), 16.
31. Thenault, *Lafayette Escadrille*, 60; André Chevrillon, quoted in Morse, *America in the War*, Part VI, Chap. XXVIII.

32. Rockwell, quoted in Morse, *America in the War*, Part VI, Chap. XXIX; Thenault, *Lafayette Escadrille*, 78.
33. Although there is a Lafayette Escadrille Memorial Cemetery located at Marnes-la-Coquette, France, Prince is not among those interred there. Rather, his remains lie at the National Cathedral in Washington, DC, a short distance from where President Woodrow Wilson is buried. American Battle Monuments Commission, Fact Sheet, https://www.abmc.gov/sites/default/files/publications/Fact%20Sheet%20Lafayette%20Escadrill%20Memorial%20List%20Rev%20Aug%202017.pdf.
34. James Rogers McConnell Memorial Collections, University of Virginia Library, https://explore.lib.virginia.edu/exhibits/show/mcconnell; James R. McConnell, *Flying for France* (Garden City, NY: Doubleday, Page, 1917), 38, 41.
35. James R. McConnell [Escadrille N 124], to Marcelle Guérin, Nice, 16 March 1917, JRM Memorial Collections, UVA Library; McConnell, *Flying for France*, 45, 65.
36. Lafayette Escadrille Casualty Report during WWI, Lafayette Escadrille Flying Corps Collection (MS 19), Clark Special Collections Branch, United States Air Force Academy; John H. Morrow Jr., *The Great War in the Air: Military Aviation from 1909–1921* (Tuscaloosa: University of Alabama Press, 1993), 336; Frandsen, *Hat in the Ring*, 272–73.
37. Thenault, *Lafayette Escadrille*, 169; Eddie Rickenbacker, *Fighting the Flying Circus*; W. David Lewis, *Eddie Rickenbacker* (Baltimore: Johns Hopkins University Press, 2005), 116; Bert Hall and James Norman Hall both survived the war and went on to publish works on their time with the Lafayette Escadrille. Hall also coauthored *Mutiny on the Bounty*.
38. Rickenbacker, *Fighting the Flying Circus*, chap. 10; *Rickenbacker: An Autobiography*, 216. Although Lufbery proved to be a more than capable trainer, he did not enjoy flying lead or flying with other pursuit aircraft at all. He told Rickenbacker he preferred to hunt alone. For this reason, his list of "official" kills is certainly smaller than his actual aerial victories. He was America's ace of aces at the time he met Rickenbacker and remained so until Rickenbacker passed him.
39. Norman Franks and Harry Dempsey, *American Aces of World War I* (Oxford, UK: Osprey Publishing, 2001).
40. Foulois, *From the Wright Brothers*, 125; Cooke, *Billy Mitchell*, 50.
41. Hurley, *Billy Mitchell*, 22–23.
42. Hurley, *Billy Mitchell*, 22–23; Cooke, *Billy Mitchell*, 66, 74–75.
43. Mason Patrick, "Final Report of the Chief of Air Service," and William C. Sherman, "Tactical History," in Maurer, *U.S. Air Service*, 1:x, 369. The original "Final Report of the Chief of the Air Service" was published in December 1918 and presented to General Pershing by the chief of the Air Service, Maj. Gen. Mason Patrick. The report was prepared by Col. Edgar S. Gorrell. The Office of Air Force History later compiled and edited the report along with a number of others into Maurer's *U.S. Air Service in World War I*. Frandsen, *Hat in the Ring*, 243, 247.
44. Rickenbacker, *Fighting the Flying Circus*, chap. 32.
45. Patrick, "Final Report," in Maurer, *U.S. Air Service*, 1:37; Frandsen, *Hat in the Ring*, 21.
46. Cooke, *Billy Mitchell*, 93–94; Patrick, "Final Report," in Maurer, *U.S. Air Service*, 1:40.

47. Foulois, *From the Wright Brothers*, 178; Hurley, *Billy Mitchell*, 36.
48. Maurer, *U.S. Air Service*, 1:40, 42.
49. Maurer, *U.S. Air Service*, 1:316; James J. Hudson, *Hostile Skies* (Syracuse, NY: Syracuse University Press, 1968), 261; Mets, *Master of Airpower*, 35.
50. Hudson, *Hostile Skies*, 262–63.
51. Maurer, *U.S. Air Service*, 1:43.
52. Patrick, "Final Report," in Maurer, *U.S. Air Service*, 1:17. By way of comparison, during World War II it took only thirty-two B-17s to drop as many pounds of bombs on a single mission.
53. Kennett, *First Air War*, 226.
54. William C. Sherman, *Air Warfare* (Maxwell AFB, AL: Air University Press, 2002), 65.
55. Mitchell, quoted in Johnny R. Jones, *William "Billy" Mitchell's Air Power* (Honolulu, HI: University Press of the Pacific, 2004), 3.
56. Paul H. Larson, "The Struggle for Control of American Military Aviation" (PhD diss., Kansas State University, 2016), 1.

Chapter 2

Mitchell, *Winged Defense*, xii.
1. William L. Mitchell, *Our Air Force: The Keystone of National Defense* (New York: E. P. Dutton, 1921), 20.
2. Brian Laslie, "Born of Insubordination: Culture, Professionalism, and Identity in the Air Arm," in Finney and Mayfield, *Redefining the Modern Military*, 203–4.
3. Douhet, *Command of the Air*, ix, 4.
4. Douhet, *Command of the Air*, 25.
5. Finney, *Air Corps Tactical School*, 16, 99; Sherman, *Air Warfare*, iii, xvii.
6. Charles F. Downs II, "Calvin Coolidge, Dwight Morrow, and the Air Commerce Act of 1926," Calvin Coolidge Presidential Foundation, 2001, https://coolidgefoundation.org/resources/essays-papers-addresses-13/. The Lampert Committee (actually the House Select Committee of Inquiry into the Operation of the United States Air Services) was convened to determine what was spent on the Air Service during the First World War, but it was certainly going to be detrimental to the administration of Calvin Coolidge. Coolidge then appointed his friend banker Dwight Morrow to head the "President's Aircraft Board."
7. Public Law 69-446, 2 July 1926; Wesley Frank Craven and James Lea Cate, *The Army Air Forces in World War II* (Washington, DC: Office of Air Force History, 1983), 1:29; Foulois, *From the Wright Brothers*, 206.
8. Foulois, *From the Wright Brothers*, 207.
9. John L. Frisbee, ed., *Makers of the United States Air Force* (Washington, DC: Office of Air Force History, 1987), 22–25; Herman S. Wolk, *Reflections on Air Force Independence* (Washington, DC: Air Force History and Museums Program, 2007), 8; Robert P. White, *Mason Patrick and the Fight for Air Service Independence* (Washington, DC: Smithsonian Institution Press, 2001), 1, 5, 7; Phillip S. Meilinger, *Airmen and Air Theory* (Maxwell AFB, AL: Air University Press, 2001), 6.
10. David E. Johnson, *Fast Tanks and Heavy Bombers* (Ithaca, NY: Cornell University Press, 1998), 158; Futrell, *Ideas, Concepts, Doctrine*, 1:68.
11. Craven and Cate, *Army Air Forces*, 1:31.

12. Joseph J. Corn, *Winged Gospel: America's Romance with Aviation* (New York: Oxford University Press, 1983), 135.
13. NARA RG 29, Records of the Department of State, Diplomatic Arrangements; NARA RG 72, Bureau of Aeronautics (Navy); NARA RG 18, Records of Army Air Forces—all in United States Army Around the World Flight (1924) Collection, National Air and Space Museum Archives, Smithsonian Online Virtual Archives, https://sova.si.edu/details/NASM.XXXX.0152; *Air Corps Newsletter*, January–June 1924, 203, and *Air Corps Newsletter*, July–December 1924, 71—both in AFHSO.
14. *Air Corps Newsletter*, January–June 1929, 16–17, AFHSO.
15. Douglas Waller, *A Question of Loyalty: Gen. Billy Mitchell and the Court-Martial That Gripped a Nation* (New York: Harper, 2004), 159; Hurley, *Billy Mitchell*, 71. Hurley refused to discuss the incident in his biography of Mitchell, but he alluded to it as something that did not need to be discussed.
16. Patrick, quoted in Cooke, *Billy Mitchell*, 133–34.
17. *Philadelphia Inquirer*, Sunday, 6 September 1925, https://www.newspapers.com/newspage/171542629/.
18. Mitchell B. J. Mulholland, "The Billy Mitchell Court-Martial of 1925" (master's thesis, University of Massachusetts, 1964), 138; Waller, *Question of Loyalty*, 325; Douglas MacArthur, *Reminiscences* (New York: McGraw-Hill, 1964), 85.
19. Carl Spaatz, OHI, K239.0512-755, 1:9, AFHRA; H. H. Arnold, *Global Mission* (New York: Harper and Brothers, 1949), 119–20.
20. Kenneth Werrell, "'Fiasco' Revisited: The Air Corps and the 1934 Air Mail Episode," *Air Power History* 57, no. 1 (Spring 2010): 12–29; Foulois, *From the Wright Brothers*, 235. The other two dates cited by Foulois were 17 December 1903, the Wright Brothers' first flight, and 19 March 1916, the 1st Aero Squadron's departure for Mexico.
21. Foulois, *From the Wright Brothers*, 238–39.
22. Foulois, *From the Wright Brothers*, 241.
23. EZAACMO Final Report, 2, AFHRA.
24. James P. Tate, *The Army and Its Air Corps* (Maxwell Air Force Base, AL: Air University Press, 1998), 144; Doolittle, quoted in Tate, 145.
25. Finney, *Air Corps Tactical School*, v.
26. Finney, *Air Corps Tactical School*, iii, 115–41.
27. Morris, *American Strategic Bombing Theory*, 172.
28. Laurence Kuter, "Growth of Air Power," unpublished manuscript, 117, 168.7012-28, AFHRA; Laslie, *Architect of Air Power*, 32. For an in-depth look at the lessons taught at ACTS and the school's instructors, see Phil Haun, *Lectures of the Air Corps Tactical School and American Strategic Bombing in World War II* (Lexington: University Press of Kentucky, 2019).
29. https://missilethreat.csis.org/wp-content/uploads/2020/09/A-Fear-for-the-Future.pdf.
30. Middlemas and Barnes, *Baldwin*, 722.
31. Middlemas and Barnes, *Baldwin*, 722.
32. Nalty, *Winged Shield, Winged Sword*, 1:142.
33. Johnson, *Fast Tanks*, 171; Kuter, "Growth of Air Power," 155.
34. Kuter, "Growth of Air Power," 155.

35. Hansell, quoted in James R. Cody, "AWPD-42 to Instant Thunder: Consistent, Evolutionary Thought or Revolutionary Change" (Maxwell AFB, AL: School of Advanced Airpower Studies, 1996), 7.
36. *Chicago Tribune*, http://archives.chicagotribune.com/1941/12/04/page/1/article/f-d-r-s-war-plans (accessed 21 March 2017).

Chapter 3

Douhet, *Command of the Air*, 10.
1. Ziemke, quoted in Earl H. Tilford Jr., *Setup: What the Air Force Did in Vietnam and Why* (Maxwell AFB, AL: Air University Press, 1991), ix.
2. William H. Bartsch, *December 8, 1941* (College Station: Texas A&M Press, 2003), 277, 281–82.
3. For a far more nuanced view of the early losses in the Pacific, see William H. Bartsch, *Every Day a Nightmare*, and *December 8, 1941*.
4. Bartsch, *December 8, 1941*, 427.
5. Johnson, *Fast Tanks*, 173; Futrell, *Ideas, Concepts, Doctrine*, 1:129.
6. James H. Doolittle, *I Could Never Be So Lucky Again* (Atglen, PA: Schiffer Publishing, 1995), 241, 274.
7. Doolittle, *I Could Never Be*, 1–2.
8. *USAAF Statistical Digest*, 112, AFHSO. For further information see Paul A. C. Koistinen, *Arsenal of World War II: The Political Economy of American Warfare, 1940–1945* (Lawrence: University Press of Kansas, 2004); Maury Klein, *A Call to Arms: Mobilizing America for World War II* (New York: Bloomsbury, 2013).
9. The most recent scholarship on the WASPs includes Katherine Sharp Landdeck, *The Women with Silver Wings* (New York: Crown, 2020); Sarah Byrn Rickman, *WASP of the Ferrying Command* (Denton: University of North Texas Press), 2016. Slightly older, but just as important, is Molly Merryman, *Clipped Wings* (New York: New York University Press, 1998); and Deborah G. Douglas, *American Women and Flight since 1940* (Lexington: University Press of Kentucky, 2004).
10. Merryman, *Clipped Wings*, 3; Reina Pennington, *Wings, Women, and War: Soviet Airwomen in World War II Combat*, 182; The GOE program supported by the Gathering of Eagles Foundation has supported this mission for more than thirty years. However, despite beginning in 1983, Jeana L. Yeager, the first woman Eagle, was not recognized until 1988. The first WASP, Dora Dougherty Strother, was recognized in 1990. http://www.goefoundation.org/.
11. Douglas, *American Women and Flight*, 85.
12. Jacqueline Cochran, *Jackie Cochran: The Autobiography of the Greatest Woman Pilot in Aviation History* (New York: Bantam Books, 1987), 210.
13. For the Women's Air Service Pilots, see Rickman, *WASP of the Ferrying Command*; and Merryman, *Clipped Wings*. For Soviet women pilots, see Pennington, *Wings, Women, and War*.
14. "Air Force Historical Program," 19 July 1942, 2, IRIS 00116419, AFHRA.
15. "Air Force Historical Program," 19 July 1942, 2, 3, IRIS 00116419, AFHRA.
16. Craven and Cate, *Army Air Forces*; Kit C. Carter and Robert Mueller, *Combat Chronology 1941–1945* (Washington, DC: Center for Air Force History, 1991); Maurer Maurer, *Combat Squadrons of the Air Force World War II* (Washington, DC: Office of Air Force History, 1982).

17. Christopher M. Rein, *The North African Air Campaign* (Lawrence: University Press of Kansas, 2012), 65–66.
18. Craven and Cate, *Army Air Forces*, 2:51, 67.
19. Case Cunningham, "William W. Momyer: An Air Power Mind," in John Andreas Olsen, *Air Commanders*, 225; Craven and Cate, *Army Air Forces*, 2:77; Momyer went on to shoot down eight aircraft, four in one engagement, and earned the Distinguished Flying Cross. He played an even larger role in airpower employment during the Vietnam War as the commander of the 7th Air Force.
20. Kuter, "Growth of Air Power," 290; Rein, *North African Air Campaign*, 133.
21. John Patrick Owens, "The Evolution of FM 100-20, Command and Employment of Air Power (21 July 1943): The Foundation of Modern Airpower Doctrine" (Master's thesis, US Air Command and General Staff College, 1975), https://apps.dtic.mil/sti/pdfs/ADA354695.pdf.
22. Owens, "Evolution of FM 100-20."
23. *Foreign Relations of the United States* [FRUS], The Conferences at Washington, 1941–42, and Casablanca, 1943, 781–82.
24. *FRUS*, Conferences at Washington and Casablanca, 781–82.
25. Alexander Fitzgerald-Black, *Eagles over Husky* (Warwick, UK: Helion & Company, 2018), xxiii.
26. Robert S. Ehlers Jr., *The Mediterranean Air War* (Lawrence: University Press of Kansas, 2015), 322.
27. Ehlers, *Mediterranean Air War*, 317.
28. James S. Corum, *Wolfram von Richthofen* (Lawrence: University Press of Kansas, 2008), 347, 353.
29. *USSBS*, 5–6.
30. James Parton, *Air Force Spoken Here*, 174; Henry King, dir., *Twelve O'Clock High*, Twentieth Century Fox Film Corporation, 1949.
31. Craven and Cate, *Army Air Forces*, 1:666.
32. Sherry, *Rise of American Air Power*, 127.
33. Haun, *Lectures*, Pointblank Directive, 267–68; Spaatz Interview, 1, Reel 30015, AFHRA.
34. Haun, *Lectures*, Pointblank Directive, 267–68.
35. Haun, *Lectures*, Pointblank Directive, 267–68.
36. Haun, *Lectures*, Pointblank Directive, 269.
37. Kuter, OHI, 1:114.
38. Donald L. Miller, *Masters of the Air* (New York: Simon and Schuster, 2006), 7.
39. Alex J. Bellamy, "Terror Bombing in the Second World War," in *Massacres and Morality: Mass Atrocities in an Age of Civilian Immunity* (Oxford: Oxford University Press, 2021), 132.
40. Craven and Cate, *Army Air Forces*, 1:661.
41. Craven and Cate, *Army Air Forces*, 1:639; The "8th Air Force" was originally designated the 8th Bomber Command on 19 January 1942 and did not transition to the new name until 22 February 1944. Organizational Records, Numbered Air Forces, 8th Air Force, AFHRA, http://www.afhra.af.mil/About-Us/Fact-Sheets/Display/Article/432272/eighth-air-force-air-forces-strategic-acc/. For simplicity's sake, I have used 8th Air Force throughout.
42. Craven and Cate, *Army Air Forces*, 1:663.

43. Laslie, *Architect of Air Power*, 68–69.
44. Parton, *Air Force Spoken Here*, 264, 385–86.
45. Craven and Cate, *Army Air Forces*, 2:632.
46. Doolittle, *I Could Never Be*, 380.
47. Spaatz, quoted in Mets, *Master of Airpower*, 194.
48. Mets, *Master of Airpower*, 197.
49. For a detailed description see Adam Tooze, *The Wages of Destruction: The Making and Breaking of the Nazi Economy* (New York: Viking Penguin, 2006); and Alfred C. Mierzejewski, *The Collapse of the German War Economy 1944–1945: Allied Air Power and the German National Railway* (Chapel Hill: University of North Carolina Press, 1988).
50. Craven and Cate, *Army Air Forces*, 3:162, 166. For an new and excellent interpretation of Allied bombing efforts in France, see Stephen Bourque, *Beyond the Beach* (Annapolis: Naval Institute Press, 2018).
51. Miller, *Masters of the Air*, 305; Mets, *Master of Airpower*, 222; Richard Davis, *Carl A. Spaatz and the Air War in Europe* (Washington, DC: Center for Air Force History, 1993), 474–76.
52. Davis, *Carl A. Spaatz*, 463.
53. *USSBS*, 37.
54. Griffith, *MacArthur's Airman*, 75, 82; Kenney, *General Kenney Reports*, 166.
55. Kenney, *General Kenney Reports*, 166.
56. Kenney, *General Kenney Reports*, 167.
57. Frederick Wesche III interview, Rutgers Oral History Archives, Rutgers University, School of Arts and Sciences, https://oralhistory.rutgers.edu/component/content/article/30-interviewees/interview-html-text/76-wesche-iii-frederick.
58. Kenney, *General Kenney Reports*, 176.
59. Kenney, *General Kenney Reports*, 205.
60. *Army Air Forces Statistical Digest*, December 1945, 307, table 211.
61. William H. Tunner, *Over the Hump* (Washington, DC: Air Force History and Museums Program, 1998), 51–52, 114.
62. Daniel Haulman, "The High Road to Tokyo Bay" (Washington, DC: Center for Air Force History, 1993), 12.
63. Herman S. Wolk, *Cataclysm: General Hap Arnold and the Defeat of Japan*, 12.
64. Marcelle Size Knaack, *Post–World War II Bombers*, vol. 2 of *Encyclopedia of US Air Force Aircraft and Missile Systems* (Washington, DC: Office of Air Force History, 1988), 494.
65. Haulman, "High Road to Tokyo," 10.
66. Werrell, *Blankets of Fire*, 135.
67. Werrell, *Blankets of Fire*, 133.
68. Werrell, *Blankets of Fire*, 138; Wolk, *Cataclysm*, 12, 120–21.
69. The 20th Air Force was the overall command for bombing operations in the Pacific. Under the 20th were the two bomber commands: XX (China) and XXI (Marianas).
70. *USSBS*, 85.
71. Kuter, OHI, 1:418.
72. *USSBS*, 92.
73. *USSBS*, 107.

74. Spaatz interview, 3, Reel 30015, AFHRA.
75. Mets, *Master of Airpower*, 302.
76. Spaatz interview, 4, Reel 30015, AFHRA.
77. Douhet, *Command of the Air*, 58; *USSBS*, 39.
78. Mark K. Wells, *Courage and Air Warfare* (London: Frank Cass, 1995), 211.

Chapter 4

Gen Laurence S. Kuter, "An Air Perspective in the Jetomic Age," *Air University Quarterly* 8, no. 2 (Spring 1956): 3.

1. Arnold to Spaatz, 6 December 1945, IRIS 01103232, AFHRA.
2. Arnold to Spaatz, 6 December 1945.
3. Mets, *Master of Airpower*, 324.
4. Walton S. Moody, *Building a Strategic Air Force*, 65.
5. LeMay, quoted in Deaile, *Always at War*, 102.
6. Letter from Arnold to Spaatz, 6 December 1945.
7. Craven and Cate, *Army Air Forces*, 1:362.
8. Vladislav M. Zubok, *A Failed Empire: The Soviet Union in the Cold War from Stalin to Gorbachev* (Chapel Hill: University of North Carolina Press, 2007), 76.
9. Daniel Harrington, *Berlin on the Brink* (Lexington: University Press of Kentucky, 2012), 296; Robert C. Owen, *Air Mobility* (Washington, DC: Potomac Books, 2013), 82.
10. John W. Leland, "Air Mobility Operations Desert Shield and Desert Storm: An Assessment," in William Head and Earl H. Tilford Jr., *The Eagle in the Desert* (Westport, CT: Praeger Press, 1996), 68.
11. RAAF, "Five Generations of Jet Fighter Aircraft," *Pathfinder* (Air Power Development Centre bulletin) 5, no. 170 (January 2012): 1.
12. Marcelle Size Knaack, *Post–World War II Fighters*, vol. 1 of *Encyclopedia of US Air Force Aircraft and Missile Systems* (Washington, DC: Office of Air Force History, 1978), 2–4.
13. Knaack, *Post–World War II Bombers*, 21.
14. USAF Fact Sheet, "B-52H Stratofortress," https://www.af.mil/About-Us/Fact-Sheets/Display/Article/104465/b-52-stratofortress/.
15. NORAD and US Northern Command Office of History, "A Brief History of NORAD," 4.
16. Reprinted in part from "The Forgotten Command," Balloons to Drones, https://balloonstodrones.wordpress.com/2016/09/22/research-note-the-forgotten-command-air-defense-command-and-the-defense-of-north-america/.
17. Lineage and honors of the ADC can be found at AFHRA: http://www.afhra.af.mil/About-Us/Fact-Sheets/Display/Article/433912/air-defense-command/.
18. NORAD began operations in September 1957. The NORAD agreement was not in effect until May 1958. NORAD is more often remembered for its annual "NORAD tracks Santa" mission or as the centerpiece of the eighties film *WarGames*.
19. NORAD and US Northern Command Office of History, "NORAD and CONAD History Reports," 1962–63.
20. Knaack, *Post–World War II Bombers*, 161.
21. Knaack, *Post–World War II Bombers*, 93.
22. Knaack, *Post–World War II Bombers*, 76.
23. Knaack, *Post–World War II Fighters*, 330.

24. Kuter, OHI, 1:522.
25. From 1997 to 2011, the USAF also had a course for newly commissioned second lieutenants: Air and Space Basic Course (ASBC), built on the USMC's Basic School model. The USAF shuttered the course in 2011.
26. Dr. Seuss, *The Butter Battle Book* (New York: Random House, 1984).

Chapter 5

Richard H. Kohn and Joseph P. Harahan, eds., *Strategic Air Warfare* (Washington, DC: Office of Air Force History, 1988), 88.

1. John Andreas Olsen, *Airpower Applied* (Annapolis: Naval Institute Press, 2017), 5; Kenneth Werrell, *Sabres over MiG Alley* (Annapolis: Naval Institute Press, 2005), 220.
2. Robert Frank Futrell, *The United States Air Force in Korea* (Washington, DC: Air Force History and Museums Program, 2000), 47; *Military Situation in the Far East: Hearings Before the Committee on Armed Services and the Committee on Foreign Relations United States Senate*, 82nd Cong., 1st sess. (1951), 3062.
3. *Military Situation in the Far East*, 3063.
4. Crane, *American Airpower Strategy*, 83.
5. 82nd Cong., "Military Situation in the Far East," 3075.
6. James Salter, *The Hunters*, 62.
7. Werrell, *Sabres over MiG Alley*, 219.
8. Crane, *American Airpower Strategy*, 53; National Museum of the United States Air Force [NMUSAF], "MiG Alley: Sabre vs. MiG," 6, https://www.nationalmuseum.af.mil/Visit/Museum-Exhibits/Fact-Sheets/Display/Article/196385/mig-alley-sabre-vs-mig/.
9. Xiaoming Zhang, *Red Wings over the Yalu* (College Station: Texas A&M Press, 2002), 201–3.
10. Gabreski had twenty-eight kills in World War II.
11. Vandenberg, quoted in NMUSAF, "Korean War Introduction," 2, https://www.nationalmuseum.af.mil/Visit/Museum-Exhibits/Fact-Sheets/Display/Article/196090/korean-war-introduction/; James T. Stewart, *Airpower* (Princeton, NJ: D. Van Nostrand, 1957), iii.
12. Alan Stephens, "The Air War in Korea 1950–1953," in Olsen, *History of Air Warfare*, 101.
13. Crane, *American Airpower Strategy*, 172, 175.
14. For command-and-control problems during the Vietnam conflict, see Wayne Thompson, *To Hanoi and Back* (Washington, DC: Air Force History and Museums Program, 2000), 15; Bernard C. Nalty, *Air War over South Vietnam, 1968–1975* (Washington, DC: Air Force History and Museums Program, 2000), 96; Jacob Van Staaveren, *Gradual Failure* (Washington, DC: Air Force History and Museums Program, 2002), 5, 15; and William Momyer, *Airpower in Three Wars* (Maxwell Air Force Base, AL: Air University Press, 2003), 92, 102.
15. Air Force Association [AFA], "The Air Force in the Vietnam War" (Arlington, VA: Aerospace Education Foundation, 2004).
16. Momyer, *Airpower in Three Wars*, 106.
17. John Schlight, *The War in South Vietnam* (Washington, DC: Air Force History and Museums Program, 1999), 289; Olsen, *History of Air Warfare*, 107–26.
18. Schlight, *War in South Vietnam*, 256–58.

19. Schlight, *War in South Vietnam*, 290.
20. Nalty, *Air War over Vietnam*, 426.
21. Wayne Thompson, "Operations over North Vietnam, 1965–1973," in Olsen, *History of Air Warfare*, 107.
22. Curtis T. Kamps, "The JCS 94-Target List," *Air and Space Power Journal* 15, no. 1 (Spring 2001): 71.
23. David M. Barrett, "Doing 'Tuesday Lunch' at Lyndon Johnson's White House," *PS: Political Science and Politics* 24, no. 4 (December 1991): 676–79.
24. Tilford, *Setup*, 109; Barrett, "Doing 'Tuesday Lunch,'" 678.
25. NMUSAF, Fact Sheets, "B-52s and Linebacker II," https://www.nationalmuseum.af.mil/Visit/Museum-Exhibits/Fact-Sheets/Display/Article/195837/b-52s-and-linebacker-ii/; Air Force Historical Studies Office, FAQs, "1972—Operation Linebacker II," https://www.afhistory.af.mil/FAQs/Fact-Sheets/Article/458991/1972-operation-linebacker-ii/.
26. Marshall Michel, "The Eleven Days of Christmas: America's Last Vietnam Battle," October 2002, http://web.mit.edu/SSP/seminars/wed_archives02fall/michel.htm.
27. Clodfelter, *Limits of Air Power*, 201.
28. Merle L. Pribbenow, ed. and trans., *Victory in Vietnam: The Official History of the People's Army of Vietnam, 1954–1975* (Lawrence: University Press of Kansas, 2002), 327.
29. Knaack, *Post–World War II Fighters*, 191, 200; AFA, "Air Force in Vietnam," 2004, 25.
30. AFA, "Air Force in Vietnam," 17, 27.
31. Mike Worden, *Rise of the Fighter Generals* (Maxwell AFB, AL: Air University Press, 1988).
32. A complete list of titles produced by the Air Force's Historical Studies Office on the Vietnam conflict can be found at http://www.afhistory.af.mil/Books/Conflicts-Wars/; a complete list of the CHECO reports can be found in Daniel S. Hoadley's thesis "What Just Happened? A Historical Evaluation of Project CHECO," School of Advanced Air and Space Studies, June 2013, https://archive.org/details/DTIC_ADA617002.

Epoch III

1. Dos Gringos, "I Wish I Had a Gun Just Like the A-10," https://www.youtube.com/watch?v=UXpyK26bwgk.
2. Laslie, *Air Force Way of War*; Worden, *Fighter Generals*; C. R. Anderegg, *Sierra Hotel* (Washington, DC: Air Force History and Museums Program, 2001); James Kitfield, *Prodigal Soldiers* (Washington, DC: Potomac Books, 1995).

Chapter 6

Gen. Robert Dixon, OHI, 1:245, AFHRA.
1. George M. Watson Jr., *Secretaries and Chiefs of Staff of the United States Air Force* (Washington, DC: Air Force History and Museums Program, 2001), 167–68.
2. Owen, *Air Mobility*, xv.
3. John W. Leland and Kathryn A. Wilcoxson, "The Chronological History of the C-5 Galaxy" (Air Mobility Command Office of History, May 2003), 1.

4. Edward T. Russell, "Military Airlift to Israel: Operation NICKEL GRASS," in A. Timothy Warnock, *Short of War* (Washington, DC: Air Force History and Museums Program, 2000), 80.
5. Air Mobility Command Office of History, "Air Refueling: Without Tankers We Cannot . . . ," October 2009, 2, 9; Laurence S. Kuter, "An Air Perspective in the Jetomic Age," *Air University Quarterly Review* 18, no. 2 (1956): 3–17, 109–23.
6. Daniel L. Haulman, "Crisis in Grenada: Operation Urgent Fury," in Warnock, *Short of War*, https://media.defense.gov/2012/Aug/23/2001330105/-1/-1/0/urgentfury.pdf, 138.
7. Haulman, "Crisis in Grenada," 138; Ronald H. Cole, "Operation Urgent Fury: Grenada" (Joint History Office, Office of the Chairman of the Joint Chiefs of Staff, 1995), 30, https://www.jcs.mil/Portals/36/Documents/History/Monographs/Urgent_Fury.pdf.
8. Haulman, "Crisis in Grenada," 141.
9. Haulman, "Crisis in Grenada," 144.
10. Ronald H. Cole, "Operation Just Cause: Panama" (Joint History Office, Office of the Chairman of the Joint Chiefs of Staff, 1995), 31, https://apps.dtic.mil/sti/pdfs/ADA313465.pdf; Just Cause, TAC Chronologies, box 1/2, folder 1, Contingencies, Air Combat Command [ACC].
11. Cole, "Operation Just Cause," 33.
12. Message from 7ACCS to 41ECS, Just Cause, TAC Chronologies, box 2/3, Contingencies, ACC.
13. Cole, "Operation Just Cause," 38–39.
14. Just Cause, TAC Chronologies, box 1/2, folder 1, Contingencies, ACC.
15. *Tactical Analysis Bulletin* [*TAB*], no. 1 (30 January 1987), ACC.
16. Joseph T. Stanik, *El Dorado Canyon* (Annapolis: Naval Institute Press, 2003), xx.
17. Many of the official after-action reports remain classified, but three good sources have been produced on this raid, including Judy C. Endicott, "Raid on Libya: Operation ELDORADO CANYON," in Warnock, *Short of War*, 145–56; Stanik, *El Dorado Canyon*; and Robert E. Venkus, *Raid on Qaddafi* (New York: St. Martin's Press, 1992).
18. 57th Wing Office of History, USAF Weapons School Fact Sheet, Nellis AFB, https://www.nellis.af.mil/About/Fact-Sheets/Display/Article/284156/united-states-air-force-weapons-school/.
19. AFR-56-3, 57th Wing Office of History, Nellis AFB.
20. AFR-56-3, 57th Wing Office of History, Nellis AFB; Weapons School lineage and honors; "Historical Brief of the Fighter Weapons School," n.d.
21. USAF Weapons School Fact Sheet, 57th Wing Office of History, Nellis AFB.
22. Boyd, quoted in Michael Hankins, "The Cult of the Lightweight Fighter: Culture and Technology in the U.S. Air Force, 1964–1991" (PhD diss., Kansas State University, 2018), 150.
23. *TAB*, no. 2 (18 April 1986): 2–3, ACC.
24. *TAB*, no. 2 (18 April 1986): 2–3.
25. *TAB*, no. 2 (18 April 1986): 2.
26. Lynn O. High, "To B(FM) or not to B(FM): That Is the Question," *TAB*, no. 2 (18 April 1986): 13–21, ACC.

Chapter 7

Glosson, *War with Iraq*, 284.

1. The "wall" of F-15s was an imperfect one. E-3 AWACS early warning aircraft detected MiGs taking off from airfields in western Iraq and committed some of the F-15s earlier than anticipated. Diane T. Putney, *Airpower Advantage* (Washington, DC: Air Force History and Museums Program, 2004), 340–42.
2. Craig Brown, *Debrief* (Atglen, PA: Schiffer Publishing, 2007), 24–25, 28.
3. "Coalition Air-to-Air Victories in Desert Storm," https://www.rjlee.org/air/ds-aakill/; Aerial Victory Tables, AFHRA. For Operation Desert Storm aerial concept of operations, see Putney's *Airpower Advantage*; and Eliot A. Cohen, ed., *Gulf War Air Power Survey* series [hereafter cited as *GWAPS*]: vols. 1–5 and *Summary Report* (Washington, DC: US Government Printing Office, 1993).
4. "Navy Training System Plan for the AIM-9M Sidewinder Missile System (for Models through AIM-9M-10)," N78-NTSP-A-50–8105C/A, May 2002, https://www.globalsecurity.org/military/library/policy/navy/ntsp/aim-9m-a_2002.pdf; Ed Rasimus, *Palace Cobra*, 21; Douglas C. Dildy and Tom Cooper, *F-15C Eagle vs MiG-23/25: Iraq 1991* (Oxford, UK: Osprey Publishing, 2016), 30, 32; *GWAPS*, 5:653–54, table 206: Coalition Air-to-Air Kill Matrix." It is important to note that Pk rates are often significantly higher in training events than during actual combat, but there is no doubt that the missiles used in Desert Storm were a vast improvement over the ones used in Vietnam.
5. Richard Davis, *On Target* (Washington, DC: Air Force History and Museums Program, 2002), 169.
6. Larry E. Cable, "Playing in the Sandbox: Doctrine, Combat, and Outcome on the Ground," in Head and Tilford, *Eagle and the Desert*, 175.
7. Hallion, *Storm over Iraq*, 83 (emphasis in original).
8. Richard T. Reynolds, *Heart of the Storm* (Maxwell AFB, AL: Air University Press, 1995), 29.
9. Buster Glosson, *War with Iraq* (Charlotte, NC: Glosson Family Foundation, 2003), 109.
10. Williamson Murray and Maj. Gen. Robert H. Scales, *The Iraq War: A Military History* (Cambridge, MA: Belknap Press of Harvard University Press, 2003), 156.
11. Mitchell, *Winged Defense*, xii.
12. Head and Tilford, *Eagle and the Desert*, 65.
13. Head and Tilford, *Eagle and the Desert*, 67–68; McPeak, quoted in Head and Tilford, 67.
14. Head and Tilford, *Eagle and the Desert*, 74; *GWAPS*, *Summary Report*, 4.
15. The nomenclature of aircraft is often called into question. For example, the F-16 is officially the F-16 Fighting Falcon, but its pilots and members of the community have long called it the Viper. The A-10 is another example, officially being named the Thunderbolt II, but more often called the Warthog, or Hog for short. In this work, I have used the name by which most members of the community would identify it.
16. Loh, quoted in John Andreas Olsen, *John Warden and the Renaissance of American Air Power* (Washington, DC: Potomac Books, 2007), 147.
17. Olsen, *John Warden*, 145.

18. John A. Warden, *The Air Campaign* (Washington, DC: National Defense University Press, 1988), xvii.
19. For the confrontation between Warden and Horner, see Reynolds, *Heart of the Storm*; Putney, *Airpower Advantage*; Tom Clancy and Chuck Horner, *Every Man a Tiger*; and "The War We Always Wanted," statement by Gen. Gilmary Hostage, quoted in Laslie, *Air Force Way of War*, 132.
20. Global Security.org, "Iraqi Air Force Equipment—Introduction," http://www.globalsecurity.org/military/world/iraq/air-force-equipment-intro.htm.
21. Putney, *Airpower Advantage*, 339–40; *GWAPS*, 5:17–20. Each SAM site had approximately six missiles on launchers and an equal number of replacements on site. For the two hundred sites, this equated to as many as 1,200 or more actual missiles ready to be launched.
22. Author interview with F-117 pilot Capt. Phil McDaniel, 23 February 2017.
23. Brown, *Debrief*, 58.
24. Author interview with F-111F pilot Maj. Gen. J. L. Briggs, 3 March 2017.
25. McDaniel interview, 23 February 2017.
26. *GWAPS*, 5:17–20.
27. The final *GWAPS* indicates a discrepancy in the weaponry coalition planners believed the Iraqis had and what they actually had. Early reports indicated that Iraq had 120 SAM batteries, but this jumped to 200 prior to the beginning of the ground portion of the campaign as more batteries were discovered through ISR and combat operations. A total of 85 SAMs remained at the conclusion of hostilities. *GWAPS*, 5:19.
28. Glosson, *War with Iraq*, 225–27.
29. 49th Wing Office of History, "Nighthawks over Iraq: A Chronology of the F-117A Stealth Fighter in Operations Desert Shield and Desert Storm," 4, https://www.f117sfa.org/nighthawk-over-iraq.
30. Williamson Murray, *Air War in the Persian Gulf* (Baltimore: Nautical and Aviation Publishing Company of America, 1995), 92. The Iraqis continued to lose aircraft even after the conflict ended. Brown, *Debrief*, 143–49.
31. BBC News, "Analysis: Iraq's Air Force," 17 March 2003, http://news.bbc.co.uk/2/hi/middle_east/2856907.stm; Operation Desert Fox Defense Department archive, http://archive.defense.gov/specials/desert_fox/ (accessed 27 February 2017).
32. David Goldfein, "A Letter from Iraq," *Combat Edge*, Spring 2012, 5–7; Aerial Victory Tables, version 2016, AFHRA (the USAF holds the official record for all USAF air-to-air combat kills). Air Force Central Command holds the lineage and honors for the former 9th Air Force.
33. *GWAPS*, 5:653–54, "Coalition Air-to-Air Kill Matrix"; Goldfein, "Letter from Iraq."
34. Aerial Victory Tables, version 2016, AFHRA.
35. Aerial Victory Tables, version 2016, AFHRA.
36. Scott A. Snook, *Friendly Fire* (Princeton, NJ: Princeton University Press, 2000), 8; Government Accountability Office, "Operation Provide Comfort: Review of U.S. Air Force Investigation of Black Hawk Fratricide," 18 January 1998, https://www.gao.gov/products/t-osi-98-13.
37. John Keegan, "Please, Mr. Blair, Never Take Such a Risk Again," *London Daily Telegraph*, 6 June 1999.

38. Brown, *Debrief*, 160–61.
39. Scott O'Grady, *Return with Honor* (New York: Doubleday, 1995), 27–28.
40. Kevin Fedarko et al., *Time*, "Rescuing Scott O'Grady: All for One," 19 June 1995, https://www.webcitation.org/5zHdPrWZu?url=http://www.time.com/time/printout/0,8816,983055,00.html (accessed 31 December 2018).
41. Lt. Col. Richard L. Sargent, "Aircraft Used in Deliberate Force," in Col. Robert C. Owen, ed., *Deliberate Force: A Case Study in Effective Air Campaign Planning* (Maxwell AFB, AL: Air University Press, 2000), 199.
42. Col. Christopher M. Campbell, "The Deliberate Force Air Campaign," in Owen, *Deliberate Force*, 109.
43. Lt. Col. Mark J. Conversino, "Executing Deliberate Force, 30 August–14 September," in Owen, *Deliberate Force*, 143.
44. Press Statement by Dr. Javier Solana, Secretary General of NATO, 23 March 1999, https://www.nato.int/docu/pr/1999/p99-040e.htm.
45. Benjamin S. Lambeth, *NATO's Air War for Kosovo* (Santa Monica, CA: RAND Corporation, 2001), 18.
46. Lt. Col. Cesar Rodriguez is a graduate of The Citadel. Rebecca Grant, "The Missing Aces," *Air Force Magazine*, 1 September 2004, http://www.airforcemag.com/MagazineArchive/Pages/2004/September%202004/0904aces.aspx.
47. "The Unthinkable, the Unimaginable Happened: An F-117 Was Shot Down in Combat," NMUSAF, https://www.nationalmuseum.af.mil/Portals/7/documents/transcripts/f117_shot_down_transcript.pdf.
48. Darrel D. Whitcomb and Forrest L. Marion, "Team Sport, Combat Search and Rescue over Serbia, 1999," *Air Power History* 61, no. 3 (Fall 2014): 32.
49. Whitcomb and Marion, "Team Sport, Combat Search," 33–34.
50. Lambeth, *NATO's Air War*, 27, 35.
51. Lambeth, *NATO's Air War*, 36–37.
52. Not only did Goldfein return to active flying status, he went on to become the twenty-first chief of staff of the Air Force in 2016. Goldfein comments from aircraft cockpit tapes: "USAF F 16 Shot Down by Serbs in 1999 Flight Data," https://www.youtube.com/watch?v=_KdmzCQxTdY.
53. Christopher E. Haave and Phil M. Haun, *A-10s over Kosovo* (Maxwell Air Force Base, AL: Air University Press, 2003), 300.
54. Daniel Haulman, "The U.S. Air Force in the Air War over Serbia," *Air Power History* 62, no. 2 (Summer 2015): 17.
55. William B. O'Conner, *Stealth Fighter: A Year in the Life of an F-117 Pilot* (Minneapolis, MN: Zenith Press, 2012), 4.

Chapter 8

https://www.kiplingsociety.co.uk/poem/poems_eastwest.htm.
1. Secretary of Defense Donald Rumsfeld announced on 25 September 2001 that the operation's name had officially changed from Operation Infinite Justice, which apparently contained too many religious overtones for the Bush administration to be comfortable with its continued use. BBC News, "Infinite Justice, Out—Enduring Freedom, In," 25 September 2001, http://news.bbc.co.uk/2/hi/americas/1563722.stm; Benjamin S. Lambeth, *Air Power against Terror* (Santa Monica, CA: RAND Corporation, 2005), 69, 92.

2. Sean Naylor, *Not a Good Day to Die* (New York: Berkley Caliber Books, 2005), 272–73.
3. Nicholas Schmidle, "Getting Bin Laden: What Happened That Night in Abbottabad," *New Yorker*, 8 August 2011, https://www.newyorker.com/magazine/2011/08/08/getting-bin-laden; Technically, Enduring Freedom ended in 2014 and was replaced by Operation Freedom's Sentinel, although the former is still commonly cited.
4. Kevin M. Woods, Mark Pease, Mark Stout, and Williamson Murray, "Iraqi Perspectives Project: A View of Operation Iraqi Freedom from Saddam's Senior Leadership," Joint Center for Operational Analysis, Joint Forces Command, January 2006, viii.
5. Woods et al., "Iraqi Perspectives Project," 31, 40.
6. NMUSAF, "Mikoyan-Gurevich MiG-25," http://www.nationalmuseum.af.mil/Visit/MuseumExhibits/FactSheets/Display/tabid/509/Article/196331/mikoyan-gurevich-mig-25.aspx.
7. Woods et al., "Iraqi Perspectives Project," 40.
8. Benjamin S. Lambeth, *The Unseen War* (Annapolis: Naval Institute Press, 2013), 63.
9. Lt. Gen. T. Michael Moseley, "Operation IRAQI FREEDOM—by the Numbers," USCENTAF, 30 April 2003; Woods et al., "Iraqi Perspectives Project," 40.
10. Moseley, "Operation IRAQI FREEDOM."
11. Lambeth, *Unseen War*, 69.
12. Author interview with Col. David Toomey, 11 May 2017.
13. Toomey interview, 11 May 2017.
14. Toomey interview, 11 May 2017.
15. Emily A. Kenney, "Bombs over Baghdad" (Office of Public Affairs, Holloman Air Force Base), 18 March 2015.
16. Toomey interview, 11 May 2017.
17. Toomey interview, 11 May 2017.
18. Woods et al., "Iraqi Perspectives Project," 128.
19. Author interview with Lt. Col. Steve Ankerstar, 24 February 2017.
20. Briggs interview, 3 March 2017.
21. Lambeth, *Unseen War*, 83.
22. Moseley, "Operation IRAQI FREEDOM." In the opening campaign of the conflict, more aircraft were lost to fratricide than to enemy action.
23. Briggs interview, 3 March 2017.
24. Operation Joint Resolve, Combined Joint Task Force, https://www.inherentresolve.mil.
25. Gen. Gilmary M. Hostage III, remarks given to the USAF First Fighter Wing, April 2012, notes in author's collection.

Chapter 9

Haave and Haun, *A-10s over Kosovo*, 300.
Michael Collins, *Carrying the Flame*, 33.
1. Thomas P. Ehrhard, *Air Force UAVs* (Arlington, VA: A Mitchell Institute Study, July 2010), 4.
2. Ehrhard, *Air Force UAVs*, 6, 10.

3. Ehrhard, *Air Force UAVs*, 25.
4. 9/11 Commission Report, 189.
5. Vandenberg, quoted in Curtis Peebles, *High Frontier* (Air Force History and Museums Program, 1997), 1.
6. Secretary of the Air Force, Letter on Space, 19 June 2018, https://www.militarynews.com/peninsula-warrior/news/air_force_news/letter-to-airmen-on-space/article_f9f80c74-74b5-11e8-b217-1b11b6506c96.html.
7. National Defense Authorization Act for Fiscal Year 2020, Conference Report to Accompany S. 1790, US House of Representatives, 116th Cong., 903–4.
8. Peebles, *High Frontier*, 11.
9. Peebles, *High Frontier*, 10.
10. Boeing Aircraft Corporation, "X-20 Dyna-Soar Space Vehicle," http://www.boeing.com/history/products/x-20-dyna-soar.page; NASA, "X-20 Dyna-Soar," https://crgis.ndc.nasa.gov/historic/X-20_Dyna-Soar (accessed 23 May 2018).
11. Barton C. Hacker and James M. Grimwood, "On the Shoulders of Titans: A History of Project Gemini," NASA Special Publication-4203, 1977, https://history.nasa.gov/SP-4203/toc.htm; Peebles, *High Frontier*, 20.
12. Peebles, *High Frontier*, 23.
13. Pat Duggins, *Final Countdown* (Gainesville: University Press of Florida, 2007), 33.
14. Global Strike Command was "bestowed" with SAC's history and heritage.
15. Official US government information about GPS can be found at https://www.gps.gov.
16. https://www.gps.gov.
17. Michael C. Whittington, "A Separate Space Force: An 80-Year-Old Argument," Air War College, Maxwell Paper No. 20 (Maxwell AFB, AL: Air War College, 2000), 1.
18. David N. Spires, *Beyond Horizons* (Colorado Springs, CO: Air Force Space Command History Office, 2007), 275.
19. FM 100-20, Command and Employment of Air Power, 21 July 1943, https://ia800208.us.archive.org/23/items/FM100_20/FM100_20.pdf.
20. "Senate, House on Collision Course over Space Corps," *USA Today*, 19 September 2017, https://www.usatoday.com/story/news/politics/2017/09/19/senate-house-collision-course-over-space-corps/682348001/.
21. The US Space Command shuttered its doors in 2002 with the stand-up of US Northern Command, so at best the new command would be a reestablished US Space Command.

Conclusion

1. Sebastian H. Lukasik, review of *Aerial Warfare: The Battle for the Skies*, by Frank Ledwidge, *Journal of Military History* 83, no. 1 (January 2019): 274–76.
2. Moffett, quoted in Air and Space Forces Association, "Quotations on Airpower," https://secure.afa.org/quotes/quotes.pdf, 15.
3. Mrozek, "Search of the Unicorn," 28–45.
4. Robin Higham and Stephen J. Harris, *Why Air Forces Fail* (Lexington: University Press of Kansas, 2006), 342.
5. Thornhill, "Over Not Through"; de Seversky, *Victory through Air Power*.

Bibliography

Albertson, Trevor. *Winning Armageddon: Curtis LeMay and Strategic Air Command, 1948–1957.* Annapolis: Naval Institute Press, 2019.
Anderegg, C. R. *Sierra Hotel: Flying Air Force Fighters in the Decade after Vietnam.* Washington, DC: Air Force History and Museums Program, 2001.
Arakaki, Leatrice R., and John R. Kuborn. *7 December 1941: The Air Force Story.* Hickam Air Force Base, HI: Pacific Air Forces, Office of History, 1991.
Arnold, H. H. *Global Mission.* New York: Harper and Brothers, 1949.
Aronstein, David C., and Albert C. Piccirillo. *Have Blue and the F-117A: Evolution of the "Stealth" Fighter.* Reston, VA: American Institute of Aeronautics and Astronautics, 1997.
Ballard, Jack S. *Development and Employment of Fixed-Wing Gunships, 1962–1972.* The United States Air Force in Southeast Asia. Washington, DC: Office of Air Force History, 1982.
Barrett, David M. "Doing 'Tuesday Lunch' at Lyndon Johnson's White House: New Archival Evidence on Vietnam Decisionmaking." *PS: Political Science and Politics* 24, no. 4 (December 1991): 676–79.
Bartsch, William H. *December 8, 1941: MacArthur's Pearl Harbor.* College Station: Texas A&M Press, 2003.
———. *Every Day a Nightmare.* College Station: Texas A&M Press, 2010.
Berger, Carl, ed. *The United States Air Force in Southeast Asia, 1961–1973: An Illustrated Account.* Washington, DC: Office of Air Force History, 1984.
Biddle, Tami D. *Rhetoric and Reality in Air Warfare: The Evolution of British and American Ideas about Strategic Bombing, 1941–1945.* Princeton, NJ: Princeton University Press, 2002.
Blesse, Frederick C. *Check Six: A Fighter Pilot Looks Back.* Mesa, AZ: Champlin Fighter Museum Press, 1987.
Bourque, Stephen. *Beyond the Beach: The Allied War against France.* Annapolis: Naval Institute Press, 2018.
Bowers, Ray L. *Tactical Airlift.* The United States Air Force in Southeast Asia. Washington, DC: Office of Air Force History, 1983.
Brown, Craig. *Debrief: A Complete History of U.S. Aerial Engagements—1981 to the Present.* Atglen, PA: Schiffer Publishing, 2007.

Burrows, William E. *This New Ocean: The Story of the First Space Age*. New York: Modern Library, 2015.
Byrd, Martha. *Chennault: Giving Wings to the Tiger*. Tuscaloosa: University of Alabama Press, 1987.
———. *Kenneth N. Walker: Airpower's Untempered Crusader*. Maxwell Air Force Base, AL: Air University Press, 1997.
Cameron, Rebecca H. *Training to Fly: Military Flight Training, 1907–1945*. Washington, DC: Air Force History and Museums Program, 1999.
Carlton, Paul K., III. "General Curtis E. LeMay on Leadership and Command." School of Advanced Air and Space Studies, Air University, June 2010.
Carter, Kit C., and Robert Mueller. *Combat Chronology 1941–1945*. 1973. Reprint, Washington, DC: Center for Air Force History, 1991.
Chertok, Boris. *Rockets and People*, vol. 1. NASA SP-2005-4110. Washington, DC: NASA History Office, 2005.
———. *Rockets and People*, vol. 2: *Creating a Rocket Industry*. NASA SP-2006-4110. Washington, DC: NASA History Office, 2006.
———. *Rockets and People*, vol. 3: *Hot Days of the Cold War*. NASA SP-2009-4110. Washington, DC: NASA History Office, 2009.
———. *Rockets and People*, vol. 4: *The Moon Race*. NASA SP-2009-4110. Washington, DC: NASA History Office, 2011.
Citino, Robert M. *The German Way of War: From the Thirty Years' War to the Third Reich*. Lawrence: University Press of Kansas, 2005.
Clancy, Tom, and Chuck Horner. *Every Man a Tiger*. New York: G. P. Putnam's Sons, 1999.
Claussen, Martin P. *Materiel Research and Development in the Army Air Arm: 1914–1945*. Army Air Force Historical Studies no. 50. AAF Historical Office, November 1946.
Clodfelter, Mark. *The Limits of Air Power: The American Bombing of North Vietnam*. New York: Free Press, 1989.
Cochran, Jacqueline. *Jackie Cochran: The Autobiography of the Greatest Woman Pilot in Aviation History*. New York: Bantam Books, 1987.
Cockburn, Andrew. *Kill Chain: The Rise of High-Tech Assassins*. New York: Henry Holt, 2015.
Cody, James R. "AWPD-42 to Instant Thunder: Consistent, Evolutionary Thought or Revolutionary Change." Maxwell AFB, AL: School of Advanced Airpower Studies, June 1996.
Cohen, Eliot A., ed. *Gulf War Air Power Survey*. 5 vols and *Summary Report*. Washington, DC: US Government Printing Office, 1993.
Cole, Ronald H. "Operation Just Cause: Panama." Joint History Office, Office of the Chairman of the Joint Chiefs of Staff, Washington, DC, 1995. https://apps.dtic.mil/sti/pdfs/ADA313465.pdf.
———. "Operation Urgent Fury." Joint History Office, Office of the Chairman of the Joint Chiefs of Staff, Washington, DC, 1997. https://www.jcs.mil/Portals/36/Documents/History/Monographs/Urgent_Fury.pdf.
Collins, Michael. *Carrying the Flame*. 40th anniversary ed. New York: Farrar, Straus and Giroux, 2009.
Cooke, James J. *Billy Mitchell*. Boulder, CO: Lynne Rienner, 2002.
———. *The U.S. Air Service in the Great War, 1917–1919*. Westport, CT: Praeger, 1996.

Corn, Joseph J., ed. *Into the Blue: American Writing on Aviation and Space Flight*. New York: Library of America, 2011.
———. *Winged Gospel: America's Romance with Aviation*. New York: Oxford University Press, 1983.
Corum, James S. *The Luftwaffe: Creating the Operational Air War, 1918–1940*. Lawrence: University Press of Kansas, 1997.
———. *Wolfram von Richthofen: Master of the German Air War*. Lawrence: University Press of Kansas, 2008.
Crane, Conrad C. *American Air Power Strategy in Korea, 1950–1953*. Lawrence: University of Kansas Press, 2000.
———. *Bombs, Cities, and Civilians: American Airpower Strategy in World War II*. Lawrence: University Press of Kansas, 1993.
Craven, Wesley Frank, and James Lea Cate, eds. *The Army Air Forces in World War II*. 7 vols. Washington, DC: Office of Air Force History, 1983. Originally published by the University of Chicago Press, 1948–58.
Cunningham, Case. "William W. Momyer: An Air Power Mind." In Olsen, *Air Commanders*.
Davies, Steve. *Red Eagles: America's Secret MiGs*. Oxford, UK: Osprey Publishing, 2008.
Davis, Burke. *The Billy Mitchell Affair*. New York: Random House, 1967.
Davis, Richard. *Bombing the European Axis Powers: A Historical Digest of the Combined Bomber Offensive, 1939–1945*. Maxwell Air Force Base, AL: Air University Press, 2006.
———. *Carl A. Spaatz and the Air War in Europe*. Washington, DC: Center for Air Force History, 1993.
———. *Hap Arnold and the Evolution of American Airpower*. Washington, DC: Smithsonian Books, 2000.
———. *On Target: Organizing and Executing the Strategic Air Campaign against Iraq*. Washington, DC: Air Force History and Museums Program, 2002.
Deaile, Melvin G. *Always at War: Organizational Culture in Strategic Air Command, 1946–1962*. Annapolis: Naval Institute Press, 2018.
De Seversky, Alexander P. *Victory through Air Power*. New York: Simon and Schuster, 1942.
Dildy, Douglas C., and Tom Cooper. *F-15C Eagle vs MiG-23/25: Iraq 1991*. Oxford, UK: Osprey Publishing, 2016.
Doolittle, James H. *I Could Never Be So Lucky Again*. Atglen, PA: Schiffer Publishing, 1995.
Douglas, Deborah G. *American Women and Flight since 1940*. Lexington: University Press of Kentucky, 2004.
Douhet, Giulio. *The Command of the Air*. Washington, DC: Office of Air Force History, 1983.
Downs, Charles F., II. "Calvin Coolidge, Dwight Morrow, and the Air Commerce Act of 1926." Calvin Coolidge Presidential Foundation, 2001, https://coolidgefoundation.org/resources/essays-papers-addresses-13/.
Duggins, Pat. *Final Countdown: NASA and the End of the Space Shuttle Program*. Gainesville: University Press of Florida, 2007.
Eastman, James N., Jr., Walter Hanak, and Lawrence J. Paszek, eds. *Aces and Aerial Victories 1965–1973*. Washington, DC: Office of Air Force History, 1976.

Echevarria, Antulio J., II. *Reconsidering the American Way of War: US Military Practice from the Revolution to Afghanistan*. Washington, DC: George Washington University Press, 2014.

Ehlers, Robert S., Jr. *The Mediterranean Air War: Airpower and Allied Victory in World War II*. Lawrence: University Press of Kansas, 2015.

Ehrhard, Thomas P. *Air Force UAVs: The Secret History*. Arlington, VA: A Mitchell Institute Study, July 2010, https://apps.dtic.mil/sti/pdfs/ADA525674.pdf.

Finney, Robert T. *History of the Air Corps Tactical School, 1920–1940*. 1955. Reprint, Washington, DC: Air Force History and Museums Program, 1998.

Fino, Steven A. *Tiger Check: Automating the US Air Force Fighter Pilot in Air-to-Air Combat, 1950–1980*. Baltimore: Johns Hopkins University Press, 2017.

Fitzgerald-Black, Alexander. *Eagles over Husky: The Allied Air Forces in the Sicilian Campaign, 14 May to 17 August 1943*. Warwick, UK: Helion & Company, 2018.

Foulois, Benjamin D. *From the Wright Brothers to the Astronauts: The Memoirs of Major General Benjamin D. Foulois*. New York: McGraw-Hill, 1968.

Frandsen, Bert. *Hat in the Ring: The Birth of American Air Power in the Great War*. Washington, DC: Smithsonian Institution Press, 2003.

Franks, Norman, and Harry Dempsey. *American Aces of World War I*. Oxford, UK: Osprey Publishing, 2001.

Frisbee, John L., ed. *Makers of the United States Air Force*. Washington, DC: Office of Air Force History, 1987.

Futrell, Robert Frank. *Ideas, Concepts, Doctrine: Basic Thinking in the United States Air Force*. Vol. 1: *1907–1960*. 1971. Reprint, Maxwell Air Force Base, AL: Air University Press, 1989.

———. *Ideas, Concepts, Doctrine: Basic Thinking in the United States Air Force*. Vol. 2: *1961–1984*. 1971. Reprint, Maxwell Air Force Base, AL: Air University Press, 1989.

———. *The United States Air Force in Korea*. Washington, DC: Air Force History and Museums Program, 2000.

———. *The Advisory Years to 1965*. The United States Air Force in Southeast Asia. Washington, DC: Office of Air Force History, 1981.

Gaston, James C. *Planning the American Air War: Four Men and Nine Days in 1941*. Washington, DC: National Defense University Press, 1982.

Glosson, Buster. *War with Iraq: Critical Lessons*. Charlotte, NC: Glosson Family Foundation, 2003.

Gordon, Michael R., and Bernard E. Trainor. *The Generals' War: The Inside Story of the Conflict in the Gulf*. New York: Little, Brown, 1995.

Gorn, Michael H. *Expanding the Envelope: Flight Research at NACA and NASA*. Lexington: University Press of Kentucky, 2001.

Grant, Rebecca. "Linebacker I." *Air Force Magazine* 95, no. 6 (June 2012).

———. "Linebacker II." *Air Force Magazine* 95, no. 12 (December 2012).

Grant, Robert L., Jr. "Vertical and Horizontal Forces: A Framework for Understanding Airpower Command and Control." School of Advanced Military Studies, Fort Leavenworth, KS, 2014.

Gray, Colin S. *Airpower for Strategic Effect*. Maxwell Air Force Base, AL: Air University Press, 2012.

Grenier, John. *The First Way of War: American War Making on the Frontier, 1607–1814*. Cambridge: Cambridge University Press, 2005.

Griffith, Charles. *The Quest: Haywood Hansell and the American Strategic Bombing in World War II*. Maxwell Air Force Base, AL: Air University Press, 1999.

Griffith, Thomas E., Jr. *MacArthur's Airman: General George C. Kenney and the War in the Southwest Pacific*. Lawrence: University of Kansas Press, 1998.

Gross, Charles. *American Military Aviation*. College Station: Texas A&M University Press, 2002.

Haave, Christopher E., and Phil M. Haun, eds. *A-10s over Kosovo: The Victory of Airpower over a Fielded Army as Told by the Airmen Who Fought in Operation Allied Force*. Maxwell Air Force Base, AL: Air University Press, 2003.

Hacker, Barton C. *American Military Technology: The Story of a Technology*. Baltimore: Johns Hopkins University Press, 2006.

Hacker, Barton C., and James M. Grimwood. "On the Shoulders of Titans: A History of Project Gemini." NASA Special Publication-4203. 1977, https://history.nasa.gov/SP-4203/toc.htm.

Hall, R. Cargill. *Case Studies in Strategic Bombardment*. Washington, DC: Air Force History and Museums Program, 1998.

Hallion, Richard. *Storm over Iraq: Air Power and the Gulf War*. Washington, DC: Smithsonian Institution Press, 1992.

Hammond, Grant T. *The Mind of War: John Boyd and American Security*. Washington, DC: Smithsonian Institution Press, 2001.

Hankins, Michael. "The Cult of the Lightweight Fighter: Culture and Technology in the U.S. Air Force, 1964–1991." PhD diss., Kansas State University, 2018.

Hansell, Haywood S., Jr. *The Air Plan That Defeated Hitler*. Atlanta: Higgins-McArthur/Longino & Porter, 1972.

———. *The Strategic Air War against Germany and Japan: A Memoir*. Washington, DC: Office of Air Force History, 1986.

Hansen, James R. *Spaceflight Revolution: NASA Langley Research Center from Sputnik to Apollo*. Washington, DC: NASA History Office, 1995.

Harrington, Daniel F. *Berlin on the Brink: The Blockade, the Airlift, and the Early Cold War*. Lexington: University Press of Kentucky, 2013.

Haulman, Daniel L. "Crisis in Grenada: Operation Urgent Fury." In Warnock, *Short of War*. https://media.defense.gov/2012/Aug/23/2001330105/-1/-1/0/urgentfury.pdf.

———. "The High Road to Tokyo Bay: The AAF in the Asiatic-Pacific Theater." Washington, DC: Center for Air Force History, 1993.

———. "The U.S. Air Force in the Air War over Serbia, 1999." *Air Power History* 62, no. 2 (Summer 2015): 6–21.

Haun, Phil. *Lectures of the Air Corps Tactical School and American Strategic Bombing in World War II*. Lexington: University Press of Kentucky, 2019.

Head, R. G. *Oswald Boelcke: Germany's Fighter Ace and Father of Air Combat*. London: Grub Street Publishing, 2016.

Head, William, and Earl H. Tilford Jr., eds. *The Eagle in the Desert: Looking Back on U.S. Involvement in the Persian Gulf War*. Westport, CT: Praeger Press, 1996.

Hennessy, Juliette. *The United States Army Air Arm, April 1861 to April 1917*. Washington, DC: Office of Air Force History, 1985.

Higham, Robin, and Stephen J. Harris, eds. *Why Air Forces Fail: The Anatomy of Defeat*. Lexington: University Press of Kentucky, 2006.

Higham, Robin, and Carol Williams, eds. *Flying Combat Aircraft of the USAAF-USAF*. Vol. 2. Manhattan, KS: Sunflower University Press, 1978.

Hoadley, Daniel S. "What Just Happened? A Historical Evaluation of Project CHECO." School of Advanced Air and Space Studies, June 2013. https://archive.org/details/DTIC_ADA617002.

Hudson, James J. *Hostile Skies: A Combat History of the American Air Service in World War I*. Syracuse, NY: Syracuse University Press, 1968.

Hughes, Thomas A. *Over Lord: General Pete Quesada and the Triumph of Tactical Air Power in World War II*. New York: Free Press, 1995.

Hughes, Thomas P. *Human-Built World: How to Think about Technology and Culture*. Chicago: University of Chicago Press, 2004.

Hurley, Alfred H. *Billy Mitchell: Crusader for Air Power*. 1964. Reprint, Bloomington: Indiana University Press, 2006.

Huston, John W., ed. *American Airpower Comes of Age: General Henry H. "Hap" Arnold's World War II Diaries*. Maxwell Air Force Base, AL: Air University Press, 2002.

Jamieson, Perry D. *Lucrative Targets: The U.S. Air Force in the Kuwaiti Theater of Operations*. Washington, DC: Air Force History and Museums Programs, 2001.

Johnson, David E. *Fast Tanks and Heavy Bombers: Innovation in the U.S. Army 1917–1945*. Ithaca, NY: Cornell University Press, 1998.

Jones, Johnny R. *William "Billy" Mitchell's Air Power*. Honolulu, HI: University Press of the Pacific, 2004.

Kaag, John, and Sarah Kreps. *Drone Warfare*. Cambridge, UK: Polity Press, 2014.

Kamps, Curtis T. "The JCS 94-Target List." *Air and Space Power Journal* 15, no. 1 (Spring 2001): 67–80.

Keaney, Thomas A., and Eliot A. Cohen. *Revolution in Warfare? Air Power in the Persian Gulf*. Annapolis: Naval Institute Press, 1995.

Kennett, Lee. *The First Air War: 1914–1918*. New York: Free Press, 1991.

Kenney, Emily A. "Bombs over Baghdad." Office of Public Affairs, Holloman Air Force Base, 18 March 2015.

Kenney, George C. *General Kenney Reports: A Personal History of the Pacific War*. Washington, DC: Office of Air Force History, 1987.

Kitfield, James. *Prodigal Soldiers: How the Generation of Officers Born of Vietnam Revolutionized the American Style of War*. Washington, DC: Brassey's, 1995.

Klein, Maury. *A Call to Arms: Mobilizing America for World War II*. New York: Bloomsbury, 2013.

Knaack, Marcelle Size. *Post–World War II Fighters, 1945–1973*. Vol. 1 of *Encyclopedia of US Air Force Aircraft and Missile Systems*. Washington, DC: Office of Air Force History, 1978.

———. *Post–World War II Bombers, 1945–1973*. Vol. 2 of *Encyclopedia of US Air Force Aircraft and Missile Systems*. Washington, DC: Office of Air Force History, 1988.

Kocher, Matthew Adam, Thomas B. Pepinsky, and Stathis N. Kalyvas. "Aerial Bombing and Counterinsurgency in the Vietnam War." *American Journal of Political Science* 55, no. 2 (April 2011): 1–18.

Kohn, Richard H., and Joseph P. Harahan, eds. *Strategic Air Warfare: An Interview with Generals Curtis E. LeMay, Leon W. Johnson, David A. Burchinal, and Jack J. Catton*. Washington, DC: Office of Air Force History, 1988.

Koistinen, Paul A. C. *Arsenal of World War II: The Political Economy of American Warfare, 1940–1945*. Lawrence: University Press of Kansas, 2004.
Kuhn, Thomas S. *The Structure of Scientific Revolutions*. 4th ed. Chicago: University of Chicago Press, 2012.
Kuter, Laurence S. "An Air Perspective in the Jetomic Age." *Air University Quarterly Review* 8, no. 2 (1956): 3–17, 109–23.
Lambeth, Benjamin S. *Air Power against Terror: America's Conduct of Operation Enduring Freedom*. Santa Monica, CA: RAND Corporation, 2005.
———. *NATO's Air War for Kosovo: A Strategic and Operational Assessment*. Santa Monica, CA: RAND Corporation, 2001.
———. *The Transformation of American Air Power*. Ithaca, NY: Cornell University Press, 2000.
———. *The Unseen War: Allied Air Power and the Takedown of Saddam Hussein*. Annapolis: Naval Institute Press, 2013.
Landdeck, Katherine Sharp. *The Women with Silver Wings: The Inspiring True Story of the Women Airforce Service Pilots of World War II*. New York: Crown, 2020.
Larson, Paul H. "The Struggle for Control of American Military Aviation." PhD diss., Kansas State University, 2016.
Laslie, Brian D. *The Air Force Way of War: U.S. Tactics and Training after Vietnam*. Lexington: University Press of Kentucky, 2015.
———. *Architect of Air Power: General Laurence S. Kuter and the Birth of the U.S. Air Force*. Lexington: University Press of Kentucky, 2017.
———. "Born of Insubordination: Culture, Professionalism, and Identity in the Air Arm." In *Redefining the Modern Military*, ed. Nathan K. Finney and Tyrell O. Mayfield (Annapolis: Naval Institute Press, 2018), 202–21.
Lavalle, A. J. C., ed. *USAF Southeast Asia Monograph Series*. Vols. 1–5. Washington, DC: Office of Air Force History, 1985.
Ledwidge, Frank. *Aerial Warfare: The Battle for the Skies*. Oxford: Oxford University Press, 2018.
Leland, John W., and Kathryn A. Wilcoxson, "The Chronological History of the C-5 Galaxy." Air Mobility Command Office of History, May 2003.
LeMay, Curtis E., and MacKinlay Kantor. *Mission with LeMay: My Story*. New York: Doubleday, 1965.
Lewis, W. David. *Eddie Rickenbacker: An American Hero in the Twentieth Century*. Baltimore: Johns Hopkins University Press, 2005.
Lukasik, Sebastian H. Review of *Aerial Warfare: The Battle for the Skies*, by Frank Ledwidge. *Journal of Military History* 83, no. 1 (January 2019): 274–76.
MacArthur, Douglas. *Reminiscences*. New York: McGraw-Hill, 1964.
Malone, Patrick M. *The Skulking Way of War: Technology and Tactics among the New England Indians*. Lanham, MD: Madison Books, 1991.
Mann, Edward. *Thunder and Lightning: Desert Storm and the Air Power Debates*. Maxwell Air Force Base, AL: Air University Press, 1995.
Mason, R. A., ed. *War in the Third Dimension: Essays in Contemporary Air Power*. London: Brassey's Defence Publishers, 1986.
Mauer, Mauer. *Aviation in the U.S. Army 1919–1939*. Washington, DC: Office of Air Force History, 1987.

———. *Combat Squadrons of the Air Force World War II*. Washington, DC: Office of Air Force History, 1982.

———, ed. *The U.S. Air Service in World War I*. Washington, DC: Office of Air Force History, 1979.

McCarthy, Gen. James P., and Col. Drue L. DeBerry, eds. *The Air Force*. Clinton, MD: Air Force Historical Foundation and Hugh Lauter Levin Associates, 2002.

McConnell, James R. *Flying for France: With the American Escadrille at Verdun*. Garden City, NY: Doubleday, Page, 1917.

McManus, John C. *Deadly Sky: The American Combat Airman in World War II*. New York: Nal Caliber, 2016.

Meilinger, Phillip S. *Airmen and Air Theory: A Review of the Sources*. Maxwell Air Force Base, AL: Air University Press, 2001.

———. *Bomber: The Formation and Early Years of Strategic Air Command*. Maxwell Air Force Base, AL: Air University Press, 2012.

———. *Limiting Risk in America's Wars: Airpower, Asymmetries, and a New Strategic Paradigm*. Annapolis: Naval Institute Press, 2017.

Merryman, Molly. *Clipped Wings: The Rise and Fall of the Women Airforce Service Pilots (WASPs) of World War II*. New York: New York University Press, 1998.

Mets, David R. *Master of Airpower: General Carl A. Spaatz*. Novato, CA: Presidio Press, 1997.

Middlemas, Keith, and John Barnes. *Baldwin: A Biography*. New York: Macmillan, 1979.

Mierzejewski, Alfred C. *The Collapse of the German War Economy 1944–1945: Allied Air Power and the German National Railway*. Chapel Hill: University of North Carolina Press, 1988.

Milett, Allan R., and Peter Maslowski. *For the Common Defense: A Military History of the United States of America*. New York: Free Press, 1984.

Miller, Donald L. *Masters of the Air: America's Bomber Boys Who Fought the Air War against Nazi Germany*. New York: Simon and Schuster, 2006.

Miller, Roger G. *Billy Mitchell: "Stormy Petrel of the Air."* Washington, DC: AFHMP, 2004.

———. *Like a Thunderbolt: The Lafayette Escadrille and the Advent of American Pursuit in World War I*. Washington, DC: Air Force History and Museums Program, 2007.

Mitchell, William L. *Our Air Force: The Keystone of National Defense*. New York: E. P. Dutton, 1921.

———. *Winged Defense: The Development and Possibilities of Modern Air Power—Economic and Military*. Mineola, NY: Dover, 2006.

Momyer, William W. *Airpower in Three Wars (WWII, Korea, Vietnam)*. Maxwell Air Force Base, AL: Air University Press, 2003.

Moody, Walton S. *Building a Strategic Air Force*. Washington, DC: Air Force History and Museums Program, 1995.

Morris, Craig F. *The Origins of American Strategic Bombing Theory*. Annapolis: Naval Institute Press, 2017.

Morrow, John H., Jr. *The Great War in the Air: Military Aviation from 1909–1921*. Tuscaloosa: University of Alabama Press, 1993.

Morse, Edwin W. *America in the War: The Vanguard of American Volunteers in the Fighting Lines and in Humanitarian Services, August 1914–April 1917*. New York: Charles Scribner's Sons, 1919.

Mortensen, Daniel R. *Airpower and Ground Armies: Essays on the Evolution of Anglo-American Air Doctrine, 1940–43*. Maxwell Air Force Base, AL: Air University Press, 1998.
Mrozek, Donald J. *Air Power and the Ground War in Vietnam: Ideas and Actions*. Maxwell Air Force Base, AL: Air University Press, 1988.
———. "In Search of the Unicorn: Military Innovation and the American Temperament." *Air University Review* 37, no. 6 (1986): 28–45.
———. *The US Air Force after Vietnam: Postwar Challenges and Potential for Responses*. Maxwell Air Force Base, AL: Air University Press, 1988.
Muehlbauer, Matthew S., and David J. Ulbrich. *Ways of War: American Military History from the Colonial Era to the Twenty-First Century*. 2nd ed. Washington, DC: Potomac Books, 2018.
Mulholland, Mitchell B. J. "The Billy Mitchell Court-Martial of 1925." Master's thesis, University of Massachusetts, 1964.
Murray, Williamson. *Air War in the Persian Gulf*. Baltimore: Nautical and Aviation Publishing Company of America, 1995.
Murray, Williamson and Maj. Gen. Robert H. Scales. *The Iraq War: A Military History*. Cambridge, MA: Belknap Press of Harvard University Press, 2003.
Nalty, Bernard C. *Air War over South Vietnam, 1965–1975*. Washington, DC: Air Force History and Museums Program, 2000.
———. *The War against Trucks: Aerial Interdiction in Southern Laos, 1968–1972*. Washington, DC: Air Force History and Museums Program, 2005.
———, ed. *Winged Shield, Winged Sword*. 2 vols. Washington, DC: Air Force History and Museums Program, 1997.
Nalty, Bernard C., John F. Shiner, and George M. Watson. *With Courage: The U.S. Army Air Force in World War II*. Washington, DC: Air Force History and Museums Program, 1994.
Naylor, Sean. *Not a Good Day to Die: The Untold Story of Operation Anaconda*. New York: Berkley Caliber Books, 2005.
Neufeld, Jacob, and George M. Watson, eds. *Coalition Air Warfare in Korea 1950–1953*. Washington, DC: Air Force History and Museums Program, 2002.
Newman, Rick, and Don Shepperd. *Bury Us Upside Down: The Misty Pilots and the Secret Battle for the Ho Chi Minh Trail*. New York: Presidio Press, 2007.
O'Brien, Phillips Payson. *How the War Was Won: Air-Sea Power and Allied Victory in World War II*. Cambridge: Cambridge University Press, 2015.
O'Conner, William B. *Stealth Fighter: A Year in the Life of an F-117 Pilot*. Minneapolis, MN: Zenith Press, 2012.
O'Grady, Scott. *Return with Honor*. New York: Doubleday, 1995.
Olsen, John Andreas, ed. *Air Commanders*. Washington, DC: Potomac Books, 2013.
———. *Airpower Applied: U.S., NATO, and Israeli Combat Experience*. Annapolis: Naval Institute Press, 2017.
———. *Airpower Reborn: The Strategic Concepts of John Warden and John Boyd*. Annapolis: Naval Institute Press, 2015.
———. *European Air Power: Challenges and Opportunities*. Washington, DC: Potomac Books, 2014.
———. *Global Air Power*. Washington, DC: Potomac Books, 2011.
———, ed. *A History of Air Warfare*. Washington, DC: Potomac Books, 2010.

———. *John Warden and the Renaissance of American Air Power.* Washington, DC: Potomac Books, 2007.
———. *Strategic Air Power in Desert Storm.* Portland, OR: Frank Cass, 2003.
Orange, Vincent. *Coningham: A Biography of Air Marshal Sir Arthur Coningham.* London: Methuen, 1990.
Overy, Richard J. *The Air War, 1939–1945.* Washington, DC: Potomac Books, 2005.
Owen, Robert C. *Air Mobility: A Brief History of the American Experience.* Washington, DC: Potomac Books, 2013.
———, ed. *Deliberate Force: A Case Study in Effective Air Campaign Planning.* Maxwell AFB, AL: Air University Press, 2000.
Pape, Robert A. *Bombing to Win: Air Power and Coercion in War.* Ithaca, NY: Cornell University Press, 1996.
Parton, James. *Air Force Spoken Here: General Ira Eaker and the Command of the Air.* 1986. Reprint, Maxwell AFB, AL: Air University Press, 2000.
Peck, Gaillard R. *America's Secret MiG Squadron: The Red Eagles of Project CONSTANT PEG.* Oxford, UK: Osprey Publishing, 2012.
Peebles, Curtis. *High Frontier: The United States Air Force and the Military Space Program.* Washington, DC: Air Force History and Museums Program, 1997.
Pennington, Reina. *Wings, Women, and War: Soviet Airwomen in World War II Combat.*
Perret, Geoffrey. *Winged Victory: The Army Air Forces in World War II.* New York: Random House, 1993.
Pribbenow, Merle L., ed. and trans. *Victory in Vietnam: The Official History of the People's Army of Vietnam, 1954–1975.* Lawrence: University Press of Kansas, 2002.
Putney, Diane T. *Airpower Advantage: Planning the Gulf War Air Campaign 1989–1991.* Washington, DC: Air Force History and Museums Program, 2004.
Quesada, Alejandro de. *The Hunt for Pancho Villa: The Columbus Raid and Pershing's Punitive Expedition 1916–1917.* Oxford, UK: Osprey Publishing, 2012.
Ramsey, Raquel, and Tricia Aurand. *Taking Flight: The Nadine Ramsey Story.* Lawrence: University Press of Kansas, 2020.
Rasimus, Ed. *Palace Cobra.* New York: St. Martin's Press, 2006.
Rayfield, Robert E. *USAF Southeast Asia Monograph Series.* Vol. 6. Washington, DC: Office of Air Force History, 1985.
Rein, Christopher M. *The North African Air Campaign: U.S. Army Forces from El Alamein to Salerno.* Lawrence: University Press of Kansas, 2012.
Reynolds, Richard T. *Heart of the Storm: The Genesis of the Air Campaign against Iraq.* Maxwell Air Force Base, AL: Air University Press, 1995.
Rich, Doris L. *Jackie Cochran: Pilot in the Fastest Lane.* Gainesville: University Press of Florida, 2007.
Rickenbacker, Eddie. *Fighting the Flying Circus.* New York: Frederick A. Stokes, 1919.
———. *Rickenbacker: An Autobiography.* Englewood Cliffs, NJ: Prentice-Hall, 1967.
Rickman, Sarah Byrn. *Nancy Love and the WASP Ferry Pilots of World War II.* Denton: University of North Texas Press, 2014.
———. *WASP of the Ferrying Command: Women Pilots, Uncommon Deeds.* Denton: University of North Texas Press, 2016.
Rood, Graham. "A Brief History of Flying Clothing." *Journal of Aeronautical History,* Paper No. 2014/01.
Salter, James. *The Hunters.* New York: Vintage Books, 1999.

Schlight, John. *Help from Above: Air Force Close Air Support of the Army, 1946–1973*. Washington, DC: Air Force History and Museums Program, 2003.

———. *The War in South Vietnam: The Years of the Offensive, 1965–1968*. Washington, DC: Air Force History and Museums Program, 1999.

———. *A War Too Long: The History of the USAF in Southeast Asia, 1961–1975*. Washington, DC: Air Force History and Museums Program, 1996.

Shaw, Robert L. *Fighter Combat: Tactics and Maneuver*. Annapolis: Naval Institute Press, 1985.

Sheehan, Neil. *A Fiery Peace in a Cold War: Bernard Schriever and the Ultimate Weapon*. New York: Random House, 2009.

Sherman, William C. *Air Warfare*. Maxwell AFB, AL: Air University Press, 2002.

Sherry, Michael S. *The Rise of American Air Power: The Creation of Armageddon*. New Haven, CT: Yale University Press, 1987.

Sherwood, John Darrell. *Officers in Flight Suits: The Story of American Air Force Fighter Pilots in the Korean War*. New York: New York University Press, 1996.

Singer, P. W. *Wired for War: The Robotics Revolution and Conflict in the 21st Century*. New York: Penguin Press, 2009.

Slife, James C. *Creech Blue: Gen Bill Creech and the Reformation of the Tactical Air Forces*. Maxwell Air Force Base, AL: Air University Press, 2008.

Snook, Scott A. *Friendly Fire: The Accidental Shootdown of U.S. Black Hawks over Northern Iraq*. Princeton, NJ: Princeton University Press, 2000.

Spires, David N. *Air Power for Patton's Army: The XIX Tactical Air Command in the Second World War*. Washington, DC: Air Force History and Museums Program, 2002.

———. *Beyond Horizons: A History of the Air Force in Space, 1947–2007*. Colorado Springs, CO: Air Force Space Command History Office, 2007.

Staaveren, Jacob Van. *Gradual Failure: The Air War over North Vietnam 1965–1966*. Washington, DC: Air Force History and Museums Program, 2002.

———. *Interdiction in Southern Laos, 1960–1968*. Washington, DC: Center for Air Force History, 1993.

Stanik, Joseph T. *El Dorado Canyon: Reagan's Undeclared War with Qaddafi*. Annapolis: Naval Institute Press, 2003.

Steiner, Barry H. *Bernard Brodie and the Foundations of American Nuclear Strategy*. Lawrence: University of Kansas Press, 1991.

Stephens, Alan, ed. *The War in the Air 1914–1994*. Maxwell AFB, AL: Air University Press, 2001.

Stewart, James T. *Airpower: The Decisive Force in Korea*. New York: D. Van Nostrand, 1957.

Tate, James P. *The Army and Its Air Corps: Army Policy toward Aviation, 1919–1941*. Maxwell Air Force Base, AL: Air University Press, 1998.

Thenault, Georges. *The Story of the Lafayette Escadrille*. Boston: Small, Maynard & Company, 1921.

Thompson, Wayne. *To Hanoi and Back: The USAF and North Vietnam, 1966–1973*. Washington, DC: Air Force History and Museums Program, 2000.

Thornhill, Paula G. *"'Over Not Through': The Search for a Strong, Unified Culture for America's Airmen."* Santa Monica, CA: RAND Corporation, 2012.

Tilford, Earl H., Jr. *Setup: What the Air Force Did in Vietnam and Why*. Maxwell AFB, AL: Air University Press, 1991.

———. *The USAF Search and Rescue in Southeast Asia*. Washington, DC: Center for Air Force History, 1992.
Tillman, Barrett. *Whirlwind: The Air War Against Japan 1942–1945*. New York: Simon and Schuster, 2010.
Toll, Ian W. *The Conquering Tide: War in the Pacific Islands, 1942–1944*. New York: W. W. Norton, 2016.
———. *Pacific Crucible: War at Sea in the Pacific, 1941–1942*. New York: W. W. Norton, 2012.
Tooze, Adam. *The Wages of Destruction: The Making and Breaking of the Nazi Economy*. New York: Viking Penguin, 2006.
Tunner, William H. *Over the Hump*. Washington, DC: Air Force History and Museums Program, 1998. Originally published in 1964.
The United States Strategic Bombing Surveys. 1945. Reprint, Maxwell AFB, AL: Air University Press, 1987.
Van Creveld, Martin. *The Age of Airpower*. New York: Public Affairs Publishing, 2011.
Venkus, Robert E. *Raid on Qaddafi: The Untold Story of History's Longest Fighter Mission by the Pilot Who Directed It*. New York: St. Martin's Press, 1992.
Waller, Douglas. *A Question of Loyalty: Gen. Billy Mitchell and the Court-Martial That Gripped a Nation*. New York: Harper, 2004.
Warden, John A. *The Air Campaign: Planning for Combat*. Washington, DC: National Defense University Press, 1988.
Warnock, A. Timothy. *Short of War: Major USAF Contingency Operations, 1947–1997*. Washington, DC: Air Force History and Museums Program, 2000.
Watson, George M., Jr. *Secretaries and Chiefs of Staff of the United States Air Force*. Washington, DC: Air Force History and Museums Program, 2001.
Weigley, Russell F. *The American Way of War: A History of the United States Military Strategy and Policy*. Bloomington: Indiana University Press, 1973.
Wells, H. G. *The War in the Air: And Particularly How Mr. Bert Smallways Fared while It Lasted*. New York: Grosset & Dunlap, 1908.
Wells, Mark K. *Courage and Air Warfare: The Allied Aircrew Experience in the Second World War*. London: Frank Cass, 1995.
Werner, Johannes. *Boelcke der Mensch, der Flieger, der Führer der deutschen Jagdfliegerei [Knight of Germany: Oswald Boelcke—German Ace]*. Leipzig: K. F. Koehler Verlag/New York: Arno Press, 1972.
Werrell, Kenneth. "The Air Force and the Future of the Strategic Bomber." *Air University Review* 27, no. 6 (1976): 73–80.
———. *Blankets of Fire: U.S. Bombers over Japan during World War II*. Washington, DC: Smithsonian Institution Press, 1996.
———. *Chasing the Silver Bullet: U.S. Air Force Weapons Development from Vietnam to Desert Storm*. Washington, DC: Smithsonian Institution Press, 2003.
———. "'Fiasco' Revisited: The Air Corps and the 1934 Air Mail Episode." *Air Power History* 57, no. 1 (Spring 2010): 12–29.
———. *Sabres over MiG Alley: The F-86 and the Battle for Air Superiority in Korea*. Annapolis: Naval Institute Press, 2005.
Whitcomb, Darrel D., and Forrest L. Marion. "Team Sport, Combat Search and Rescue over Serbia, 1999." *Air Power History* 61, no. 3 (Fall 2014) 28–37.

White, Robert P. *Mason Patrick and the Fight for Air Service Independence.* Washington, DC: Smithsonian Institution Press, 2001.
White, William D. *U.S. Tactical Air Power: Missions, Forces, and Costs.* Washington, DC: Brookings Institution, 1974.
Whittington, Michael C. "A Separate Space Force: An 80-Year-Old Argument." Air War College, Maxwell Paper No. 20. Maxwell AFB, AL: Air War College, 2000.
Whittle, Richard. *Predator: The Secret Origins of the Drone Revolution.* New York: Henry Holt, 2014.
Wolk, Herman S. *Cataclysm: General Hap Arnold and the Defeat of Japan.* Denton: University of North Texas Press, 2010.
———. *Planning and Organizing the Postwar Air Force, 1943–1947.* Washington, DC: Air Force History and Museums Program, 1984.
———. *Reflections on Air Force Independence.* Washington, DC: Air Force History and Museums Program, 2007.
———. *The Struggle for Air Force Independence, 1943–1947.* Washington, DC: Air Force History and Museums Program, 1997.
Woods, Kevin M., Mark Pease, Mark Stout, and Williamson Murray. "Iraqi Perspectives Project: A View of Operation Iraqi Freedom from Saddam's Senior Leadership." Joint Center for Operational Analysis, Joint Forces Command, January 2006.
Worden, Mike. *Rise of the Fighter Generals: The Problem of Air Force Leadership, 1945–1982.* Maxwell Air Force Base, AL: Air University Press, 1998.
Y'Blood, William T., ed. *The Three Wars of Lt. Gen. George E. Stratemeyer: His Korean War Diary.* Washington, DC: Air Force History and Museums Program, 1999.
Zhang, Xiaoming. *Red Wings over the Yalu: China, the Soviet Union, and the Air War in Korea.* College Station: Texas A&M Press, 2002.
Zubok, Vladislav M. *A Failed Empire: The Soviet Union in the Cold War from Stalin to Gorbachev.* Chapel Hill: University of North Carolina Press, 2007.

Index

Ace Drummond (comic strip), 46–47
aerial bombardment. *See* strategic bombardment
Aerial Demonstration Team (Thunderbirds), 106
aerial reconnaissance: about development and role, 1, 2, 8, 200; beginning in Mexican Expedition, 29–30, 200; early uses in WWI, 30–31, 45–46, 54; execution in WWII, 69–70, 74, 78; support of Desert Storm, 155–56; use in Afghanistan, 175; use in Grenada, 137; use of UAVs in, 190–91; uses in space including satellites, 194–95
aerial refueling: capability for, 1, 2; concept tested (the "Question Mark" flight), 54, 65, 136; developing technology and training for, 49, 100, 142, 148; introduction of C-141 Starlifter, 134–35; introduction of KC-135 Stratotanker, 136; projecting global air power, 135; support of Desert Storm, 155–56; support of El Dorado Canyon, 140; support of Enduring Freedom, 175–76; support of Iraqi Freedom, 180–81
aerial tactics and doctrine, 139; angle of attack, 142; AWPD-1, 63–65, 67–70, 91, 104; basic fighter maneuvers (BFM), 142–46; Boelcke's "Dicta," 32; box formation (bombers), 85; "boxing in" (fighters), 18; Combat Archer/Combat Hammer exercises, 131–32, 139, 152, 162; Combined Bomber Offensive (CBO), 77–85, 89; "Command and Employment of Air Power" (FM 100-20), 79–80, 104, 198; Cope Thunder exercises, 185; energy maneuverability theory (EMT), 143–44; "five rings" (target selection methodology), 182–83, 208n15; as "good stick," 118; high-altitude daylight precision bombing (HADPB), 9, 69, 84, 94; look down–shoot down, 163; low-altitude bombing with incendiaries (firebombing), 95–96; "MiG trains," 18; nighttime bombing, 44, 58, 84, 95–96, 126, 158–59; Pointblank directive, 82–83, 86; radius/rate of turn, 142; six o'clock position, 118; skip bombing, 14, 91; *Tactical Analysis Bulletin*, 144–46; "3-9 Line," 142–43; training exercises (Red Flag), 21, 140–42, 144, 148, 152, 161–63, 185
Aerial Warfare (Ledwidge), 4
Afghanistan. *See* Operation Enduring Freedom
airborne warning and control system (AWACS), 21, 163–64, 166, 169, 222n1
Air Campaign, The (Warden), 157
Air Combat Command (ACC), 22, 158, 196, 209n33
Air Commerce Act of 1926, 51–52
Air Corps Service School, 54–55

Air Corps Tactical School (ACTS, formerly Air Service Tactical School), 8–10, 51, 60–64, 82–84, 94, 208n13

aircraft (attack): A-6 Intruders, 149; A-7 Corsairs, 127, 148, 149; A-10 Thunderbolt II (aka "Warthog"), 21, 137, 155, 183–84, 189, 201, 222n15; A-20 Havoc, 91

aircraft (bombers): B-1 Lancer, 175, 183–84; B-2 Spirit, 171–72, 175, 177, 179; B-10 (Martin B-10), 62; B-17 Flying Fortress, 62, 67, 72, 81–82, 84, 91–92, 94, 106; B-18 Bolo, 61; B-24 Liberator, 62, 67, 77, 84, 85, 94; B-25 Mitchell Bomber, 73–74, 91; B-26 Marauder, 91; B-29 Superfortress, 13, 16, 17, 19, 67, 94–96, 106, 114–15; B-36 Peacemaker, 16, 19, 68, 106–7, 136; B-45 Tornado, 110; B-47 Stratojet, 16, 19, 68, 107, 132, 136; B-50 Superfortress, 19, 110; B-52 Stratofortress, 16; XB-70 Valkyrie, 110; YB-9, 62

aircraft (fighters): about the Century Series, 16, 109; F-4 Phantom II (aka "Wild Weasel"), 128, 150, 153, 155, 161, 191; F-5 Tiger, 141, 146; F-14 Tomcat, 149; F-15 Eagle, 21, 131–32, 137–38, 149–50, 154, 159, 161, 163–64, 169, 173, 176, 183–84, 210n38, 222n1; F-16 Fighting Falcon (aka "Viper"), 21, 131, 143, 155, 163, 165–66, 169, 171, 173, 176, 181, 183–84, 201, 222n15; F-18 Hornet, 149; F-22 Raptor, 105, 173, 177, 201, 203; F-35 Lightning II, 22, 105, 173, 183, 201; F-51 Mustang, 134; F-80 Shooting Star (aka P-80), 105–6; F-84 Thunderjet, 106; F-100 Super Sabre (aka "Hun"), 109, 122, 127, 142; F-101 Voodoo, 109; F-102 Delta Dagger, 109; F-104 Starfighter, 109, 127; F-105 Thunderchief, 109, 128; F-106 Delta Dart, 109; F-108 Rapier, 110–11; F-111 Aardvark, 128, 132, 140, 155, 159; F-117 Nighthawk, 132, 138–39, 147–48, 149, 155, 158–59, 161, 169, 170, 172, 179–82, 202; CF-105 Arrow (Canadian), 111; P-26 Peashooter, 62–63; P-35 (Seversky), 63, 73; P-36 Hawk, 63; P-40 Warhawk, 72, 78; P-47 Thunderbolt, 13; P-51 Mustang, 13, 83; P-80 Shooting Star (aka F-80), 105–6. *See also* pursuit aviation

aircraft (other): Albatross (German WWI fighter), 38; Fokker C-2A (German trimotor), 54; Fokker Dr.1 (German triplane), 26, 38, 43; Gotha (German heavy bomber), 33–34; LVG B (German reconnaissance plane), 37; Me-262 (German fighter), 105; Mirage F-1 (French fighter), 150; Nieuport 11 (French fighter, aka "Bébé"), 37; Spads (French fighter), 43; Zeppelin (German airships), 31, 33–34

aircraft (other US): C-5 Galaxy ("FRED"), 135–36, 137–38, 155; AH-64 Apache (helicopter), 149, 170; C-17 Globemaster III, 136; C-54 Skymaster, 105; C-130 Hercules, 136–38; C-141 Starlifter, 134–36, 137–38, 155; E-3 Sentry (AWACS), 21, 163–64, 166, 169, 222n1; E-8 Joint STARS (AWACS), 21; EA-6B Prowler (electronic warfare platform), 181; EF-111 Raven, 150; EF-111 Raven, 138, 150; HB1001/1002 Have Blue, 147; JN-3/JN-4 (Curtis "Jenny"), 29–30; KC-10 Extender, 136, 138, 155; KC-135 Stratotanker, 136, 138, 155, 181, 191; KC-46 Pegasus, 22, 137; MC-130 Combat Talon, 137; MH-47 (search and rescue helicopter), 155; MH-53 Pave Low (special operations helicopter), 149, 170; MH-60 (search and rescue helicopter), 155; O-1 Bird Dogs, 120; T-38 Talon, 141, 142; U-2 Dragon Lady, 137; UH-1 Iroquois ("Huey"), 191; UH-60 Black Hawk (helicopter), 163–64; Wright Flyer (Signal Corps No. 1), 28–29, 200. *See also* unmanned aerial vehicles/systems

INDEX 243

aircraft (Soviet): Mi-24 Hind (helicopter), 163; MiG-15 Fagot, 17–19, 106, 116–19, 121, 124; MiG-21 Fishbed, 141, 150, 161; MiG-25 Foxbat, 158–59, 163, 178; MiG-29 Fulcrum, 150, 161, 169; SU-22 (aka SU-17, "Fitter"), 163

aircraft (UAVs and space): MQ-1 Predator, 187, 191; RQ-1 Predator, 191; satellites and manned space flight, 195–96; X-20 Dyna-Soar (space plane), 195

aircraft and crew losses: Iraq, 159–61, 225n22; Korea, 18, 118; UAVs, 191; Vietnam, 123, 126, 128; WWII (American), 14, 71–72, 81–83, 93, 98–99; WWII (British), 81; WWII (German), 86–88; Yugoslavia, 169, 172

aircraft technology and munitions: AESA radar, 131; AGM-88 HARM missiles, 150; AGM-114 Hellfire missile, 191; AIM air-to-air missiles, 131; AIM heat-seeking missiles, 150–51; AIR-2 Genie nuclear rocket, 109; AN/APG-63 radar, 131; Bomarc missiles, 109; electronic warfare, 126, 131, 150, 190; GBU-27 bombs, 149, 180–81; GPS and weapons guidance, 197–98; "hands on throttle and stick" (HOTAS), 150–51; head-up display, 151; low observable (stealth) technology, 147–48, 161, 172, 181; M61A1 rotary cannon, 131; Nike and Zeus surface-to-air missiles, 109; precision-guided munitions (PGM), 125, 140, 160–62, 171–73; "Spark Varks" (electronic jammers), 150; terrain following radar (TFR), 159; UAVs and space systems in, 187–88

aircraft units: 1st Aero Squadron, 29, 214n20; 1st American Pursuit Group, 45; 1st Bombardment Wing, 85; 1st Provisional Air Brigade, 55; 1st Tactical Fighter Wing, 154; 2nd Air Division, 120; 2nd Air Force, 77; 2nd American Pursuit Group, 45; 3rd American Pursuit Group, 45; 7th Air Force, 20–21, 120–21, 216n19; 8th Air Force, 70, 77, 80, 83–86, 88–89, 95, 126, 216n41; 8th Bomber Command, 216n41; 8th Fighter Command, 83, 86–87; 9th Air Force, 88, 223n32; 15th Air Force, 88, 115; 20th Air Force, 92–96, 196–97, 217n69; 27th Aero Squadron, 26; 33rd Fighter Group, 78; 33rd Tactical Fighter Wing, 138; 36th Tactical Fighter Wing, 154; 37th Tactical Fighter Wing, 161; 47th Pursuit Squadron, 71; 48th Tactical Fighter Wing, 139–40; 94th Aero Squadron, 25, 40; 94th Fighter Squadron, 25; 509th Composite Bomb Group, 97; 555th Fighter Squadron, 170; Aggressor squadrons, 141, 146, 148; Task Force 77 (US Navy), 121; XII Bomber Command, 82; XX Bomber Command, 95; XXI Bomber Command, 92, 95

air defense: about components of IADS, 158; bombardment doctrine and, 12–13, 18–19, 20, 95–96; countering threat of, 140, 170, 172; creating air superiority over, 150, 158–60, 162–63, 167–68, 178–79, 184–85; intercepting incoming aerial attack, 107–9, 124; interceptor aircraft designed for, 110–11, 131

Air Defense Command (ADC), 16–17, 102–3, 107, 108–9, 112–13. *See also* Continental Air Defense Command; North American Air Defense Command

Air Force, The (ed. McCarthy and DeBerry), 4

Air Force Historical Foundation, 121

Air Force Historical Studies Office, 126, 220n32

Air Force History and Museums Program, 129

Air Force of South Vietnam (VNAF), 122

Air Force Space Command (AFSPACE), 22, 193, 197–98

Air Force Way of War, The (Laslie), 3, 4

Air Mail scandal of 1934, 57–59, 65

air mobility, 134–36

Air Power and Armies (Slessor), 51
airpower and "way of war." *See* "way of war"
Airpower for Strategic Effect (Gray), 3
Airpower: The Decisive Force in Korea (Stewart), 119
Air Service Newsletter, 54
air superiority: air-to-air combat to gain, 18; developing doctrine in WWI, 31, 42–47, 49; developing doctrine in WWII, 64–65, 70; development of aircraft for, 131, 149, 210n38; kill-ratios, 18, 210n39; need for, 22; operations and achievement in WWII, 86–88; operations in Desert Shield, 155–56, 158–59; operations in Desert Storm, 150–51, 154; operations in Korea, 118–19; unlearning lessons of, 90, 119–20. *See also* pursuit aviation
air support (military operations): 1942, North Africa, invasion of (Operation Torch), 77; 1943, German factories, bombing of (Operation Pointblank), 85–86; 1943, Italy, invasion of (Operation Avalanche), 80–81; 1943, Ploesti, raid on (Operation Tidal Wave), 85–86; 1943, Tunisian campaign (Operation Flax), 78; 1944, bombing of Japan (Operation Matterhorn), 93; 1944, breakout from Normandy (Operation Cobra), 88–89; 1944, Germany, achieving air superiority over (Operation Argument), 86–88; 1944, Normandy landings (Operation Overlord), 86; 1945, Japan, firebombing of (Operation Meetinghouse), 95–96; 1945, Japan, mining harbors of (Operation Starvation), 95–96; 1948, Berlin Airlift (Operation Vittles), 103–5; 1951, Korean conflict (Operation Strangle I and II), 116; 1952, Korean conflict (Operation Saturate), 116; 1964, Vietnam War (Operation Barrel Roll), 125; 1965, Vietnam War (Operation Rolling Thunder), 123–25; 1965, Vietnam War (Operation Steel Tiger), 125; 1967, Vietnam War (Operation Cedar Falls), 121; 1967, Vietnam War (Operation Junction City), 121; 1967, Vietnam War (Operation Linebacker I and II), 123, 125–27; 1973, Yom Kippur War (Operation Nickel Grass), 135, 177; 1980, Iran hostage rescue attempt (Operation Eagle Claw), 148; 1983, Bosnia and Herzegovina, intervention in (Operation Deny Flight), 22, 164–66; 1983, Grenada, invasion of (Operation Urgent Fury), 133–34, 137–39, 148; 1986, Libya, bombing of (Operation El Dorado Canyon), 133–34, 137, 139–40, 148, 221n17; 1989, Panama, invasion of (Operation Just Cause, formerly Blue Spoon), 133–34, 137–39, 148; 1990, Gulf War (Operation Desert Shield), 135, 154–55, 177; 1990, Gulf War (Operation Instant Thunder), 156–58; 1991, Gulf War (Operation Desert Storm), 14, 21–22, 107, 118, 135, 148–56, 159–63, 167, 172–79, 183–85, 201–2, 222n4; 1992, Iraq, monitoring Iraqi air space (Operation Southern Watch), 162–64, 177; 1995, Republica Srpska, intervention in (Operation Deliberate Force), 22, 164–65, 167–68; 1997, Iraq, enforcing no-fly zone (Operation Northern Watch), 162–64, 177; 1998, Iraq, bombing of (Operation Desert Fox), 162; 1999, Yugoslavia, bombing of (Operation Allied Force, aka Noble Anvil), 22, 107, 118, 164–65, 168–73, 187; 2001, Afghanistan, war in (Operation Enduring Freedom, formerly Infinite Justice), 107, 174–77, 184, 191, 201, 224n1, 225n3; 2002, Battle of Shahi Kot (Operation Anaconda), 176–77; 2002, Iraq, attacking air defenses in (Operation Southern Focus), 178–79; 2003, Iraq War (Operation Iraqi

INDEX   245

Freedom), 107, 153, 162, 174, 177–85, 191–92; 2014, Syria, destruction of ISIS in (Operation Inherent Resolve), 184; 2015, continuation of Enduring Freedom (Operation Freedom's Sentinel), 225n1; 2018, Islamic State, war in Syria against (Operation Inherent Resolve), 192, 198, 200
air support (US Marine Corps), 94, 116, 119, 149, 152, 156
air support (US Navy), 14, 29, 77–78, 94, 116, 125, 140, 149–50, 156, 175
Air Transport Command (ATC), 101. *See also* Military Air Transport Service
Air University (AU), 112
Air Warfare (Sherman), 47, 50–51
Aldrin, Edwin E., Jr. ("Buzz"), 117
Al Udeid Air Base (Qatar), 175, 180
America in the War: The Vanguard of American Volunteers (Morse), 35
American Airpower Strategy in Korea (Crane), 16
American Air Service (AEF air arm), 27, 34, 39–46, 202
American Expeditionary Force (AEF). *See* American Air Service
American Military Aviation (Gross), 4
American Military Technology (Hacker), 12–13
American Way of War (Weigley), 3
Andrews, Frank M., 53, 60
Ankerstar, Steven ("Cruiser"), 182
Area 51 (Nevada), 148
Armstrong, Frank A., 81
Armstrong, Neil, 195
Army Air Defense Command, 112–13
Army Air Forces in World War II, The (Craven and Cate), 52, 77
Army Antiaircraft Command (ARAACOM), 198
Arnold, Eleanor Pool ("Bee"), 102
Arnold, Henry Harley ("Hap"): Air Force history, preserving, 76; Air Force leadership role, 19, 63, 65, 70; Air Mail scandal of 1934, role in, 58; bombing campaign strategy, 13, 80, 82, 84, 94–95, 97; Mitchell court martial, 57; retirement and leadership transition, 101–3
A-10s over Kosovo (Haave and Haun), 189
atomic bomb/atomic weapons: competition with bombardment doctrine, 112; focus on delivery systems, 67–68, 99, 106–7, 110; impact on warfare, 14–16, 100–102; reexamining use against Japan, 96–97
AWPD-1 (air war plan for WWII), 63–65, 69–70, 91, 104

Baker, Newton, 59
Baldwin, Stanley, 9, 61–62
Balkan air campaigns (1990s), 22, 81, 152–54, 162, 164–73, 191, 201
Balsley, Clyde, 38, 39
Bartsch, William H., 71, 72
Battle of Château-Thierry (1918), 44
Battle of Khe Sanh (1968), 160
Battle of Saint-Mihiel (1918), 35, 43–44
Battle of the Bismarck Sea (1942), 92
Battle of the Marne (1914), 31
Battle of the Meuse-Argonne (1918), 44–46
Battle of Tora Bora (2001), 176
Beard, Charles, 11
Bellows Army Air Field (Hawaii), 58
Berlin Airlift, 93, 103–5, 135, 155, 177
Berlin on the Brink (Harrington), 105
Biddle, Charles, 45
Bishop, Maurice, 137
Bishop, William ("Willy"), 31
Blesse, Frederick C.("Boots"), 118
Board of Aeronautic Inquiry (aka Morrow Board), 52
Boelcke, Oswald, 32, 34, 211n14
Boelcke's "Dicta" (rules for fighting in the air), 32, 34
bombardment aviation. *See* strategic bombardment
Borman, Frank, 194
Bosnia and Herzegovina, 22, 165–68
Boyd, John ("Forty-Second Boyd"), 21, 131, 133, 141, 143–44, 181

Bradley, Omar, 88
Branch, Harlee, 58
Braun, Wernher von, 194
Brereton, Lewis Hyde, 71–72
Brett, George Howard, 60, 70
Briggs, J. L., 159, 182–83
Brown, David, 11
Brown, Lord Stanley, 208n14
Bush, George H. W., 138
Bush, George W., 177, 224n1
Butter Battle Book (Geisel), 112
Byrd, Martha, 10

Cable, Larry, 153
Campbell, Christopher M., 167
Canada, 31, 108, 111, 149
Cannon, Joseph Gurney ("Joe"), 28–29
Cardoso, Jim, 170
Carlton, Paul K., 135
Carrying the Flame (Collins), 189
Cate, James Lee, 52
Chapman, Victor, 36, 38–39
Cheney, Richard ("Dick"), 139
Chennault, Claire, 8–10, 49, 60–61, 202, 208n13
Chennault: Giving Wings to the Tiger (Byrd), 10
Chevrillon, André, 38
Cheyenne Mountain Air Force Station, 108
Chidlaw, Benjamin, 108
circumnavigation of the globe, 49, 54
Civil Aeronautics Authority Act of 1938, 51–52
Clark, Wesley K., 168, 170
Clausewitz, Carl von, 10, 159
Clinton, William J. ("Bill"), 168, 172
Clodfelter, Mark, 21, 123, 127
close air support (CAS): about pre-WWII doctrine for, 64–65, 69–70; aircraft adaptations to, 100–101, 115, 131, 174, 201; bombers use in, 88–89, 107, 183; deficiencies in Afghanistan, 176–77; deficiencies in Korea, 17, 18–19; friendly-fire incidents, 89; operations and achievements in WWII, 43–44, 77, 99; operations in Desert Storm, 154; operations in Grenada, 137; operations in Vietnam, 120–21; UAVs and, 175
Cochran, Jacqueline ("Jackie"), 74–75, 202
Collins, Michael, 189
Combat Chronology, 1941–1945 (Carter and Mueller), 77
Combat Squadrons of the Air Force in World War II (Mauer), 77
Combined Bomber Offensive (CBO), 77–85, 89
Combined Joint Task Force (CJTF). *See* Operation Inherent Resolve
Command of the Air (Douhet), 50, 69, 208n15
Contemporary Historical Evaluation of Combat Operations (CHECO reports), 129, 220n32
Continental Air Defense Command (CONAD), 108–9, 111
Coolidge, Calvin, 213n6
Cooper, Leroy ("Gordo"), 194
Corbett, Julian, 10
Corn, Joseph, 54
Corum, James, 81
Cowdin, Elliott, 35–36
Crane, Conrad, 16–18, 116, 118, 119
Craven, Wesley Frank, 52
Cripen, Bob, 196
Cummings, Homer, 58

Dani, Zoltán, 169
Davis, Richard, 89, 152
DeBellevue, Charles ("Chuck"), 128
DeBerry, Drue L., 4
Defense Advanced Research Projects Agency (DARPA), 147–48
Defense Department Reorganization Act of 1986, 156
Deptula, David, 157
Dern, George, 58
De Seversky, Alexander P., 5–6, 11, 63, 82, 135–36, 173
Disney, Walt, 12
Doolittle, James Harold ("Jimmy"), 59, 65, 77, 83, 86, 90

Doolittle Raid, 73–74, 90–91, 94
Douhet, Giulio, 9, 11, 19, 33–34, 50, 63, 69, 203
drones. *See* unmanned aerial vehicles/systems (UAVs/UAS)
Drum, Hugh Aloysius, 42, 53, 58–59
Dugan, Mike, 156
Dunford, Joseph, 193

Eagle in the Desert, The (Head), 154–55
Eagles over Husky (Fitzgerald-Black), 80
Eaker, Ira, 49, 59–60, 65, 73, 81
"Early History of Army Aviation" (Burge), 28–29
Echevarria, Antulio J., II, 2, 3–4
Ehlers, Robert S., 81
Ehrhard, Thomas, 190
Eisenhower, Dwight D., 16, 89, 194, 196
Emmons, Delos C., 53
Ent Air Force Base (Colorado), 108
"Escuadrille Americaine" (aka Lafayette Flying Corps). *See* Lafayette Escadrille

Fairchild, Muir S. ("Santy"), 8, 60–61, 111
Far East Air Force (FEAF), 17, 115
Farley, James, 58
Fechet, James, 51, 52, 65
Feinstein, Jeff, 128
Fighting in the Air (McCudden), 32
Fighting the Flying Circus (Rickenbacker), 43
"Final Report of Chief of Air Service" (Patrick), 46, 212n43
First Air War: 1914–1918, The (Kennett), 8, 19, 46
First Way of War, The (Grenier), 4
Fitzgerald-Black, Alexander, 80
Flying for France (McConnell), 39
"flying the Hump," 92–93, 177
Flying Tigers (aka First American Volunteer Group), 10
Fogelman, Ronald, 166
Fonck, René, 31
Foulois, Benjamin: airpower doctrine developed/defined, 34; involvement in Air Mail scandal, 57–59; leadership in early US aviation, 7, 28–29, 47, 51–53, 65; Mexican Expedition aviation support (1916-17), 29–30, 214n20; relationship with Mitchell, 41–42, 44
Fox, Mark, 150
Frandsen, Bert, 38
French Aviation Service, 35
French Foreign Legion, 35, 37
Friendly Fire: The Accidental Shootdown (Snook), 164
From the Earth to the Moon (Verne), 194
From the Wright Brothers to the Astronauts (Foulois), 29
Futrell, Robert Frank, 4

Gabreski, Francis S. ("Gabby"), 118, 219n10
Gabriel, Charles A., 21, 128, 134, 210n43
Gaddhafi, Mu'ammar, 139
Gathering of Eagles (GOE), 215n9
Geczy, Michael H., 169
Geisel, Theodor Seuss (aka Dr. Seuss), 112
General Headquarters Air Force (GHQ Air Force), 53
Geographic Combatant Commands (GCC), 198
George, Harold Lee, 59, 60, 63
German Way of War, The (Citino), 4
Giffard, Hardinge Goulburn (2nd Earl of Halsbury, aka Lord Tiverton), 33
Glenn, John H., Jr., 117
Global Mission (Arnold), 102
Global Positioning System (GPS), 181, 191, 197–98, 226n15
Global Strike Command (GSC), 22–23, 197, 226n14
Glosson, Buster Cleveland, 153, 157, 161
Goddard, Robert H., 194
Goldfein, David, 162–63, 170–71, 224n52
Goldwater-Nichols Act of 1986, 156
Gorrell, Edgar, 34, 59, 212n43
Graeter, Rob, 150
Gray, Colin S., 1, 3
Grenada. *See* Operation Urgent Fury
Griffith, Thomas E., 14

Grissom, Virgil I. ("Gus"), 117, 194
Gross, Charles, 4
Gulf War. *See* Operation Desert Shield; Operation Desert Storm
Gulf War Air Power Survey (GWAPS), 158, 179, 223n27

Haave, Christopher E., 189
Hacker, Barton, 12
Hall, James Norman, 40, 212n37
Hall, Weston Birch ("Bert"), 36, 38, 212n37
Hallion, Richard, 1, 60, 153
Halvorsen, Gail S. ("Candy Bomber"), 105
Hansell, Haywood, 60, 63, 65, 70, 85, 91, 94–95
Hardin, Thomas O., 93
Harmon, Millard F., 76
Harrington, Daniel, 105
Harris, Stephen, 203
Haulman, Daniel, 93
Haun, Phil M., 189
Head, William, 154–55
Hennessy, Juliette, 34
Hickam, Horace, 58
Hickam Army Air Field (Hawaii), 70–72
High, Lynn O., 146
Higham, Robert, 203
Hinote, Clint, 180
History of Air Warfare, A (Olsen), 4, 121
Hoehn, Mark, 180–81
Holloman Air Force Base (New Mexico), 180
Horner, Charles ("Chuck"), 152–53, 156–58, 203
Hughes, Thomas, 10, 13, 15
Humphreys, Frederick E., 29
Hunters, The (Salter), 117
Hussein, Saddam, 153–54, 174, 177–78, 180
Hwang, Jeff, 169

I Could Never Be So Lucky Again (Doolittle), 73
Ideas, Concepts, Doctrine: Basic Thinking (Futrell), 4

insurgency/counterinsurgency operations, 29, 175, 183–84
integrated air defense systems (IADS). *See* air defense
intercontinental ballistic missiles (ICBMs), 22–23, 103, 196–97
Iran hostage crisis, 137–38, 148
Iran-Iraq War, 180
Iraq. *See* Operation Desert Fox; Operation Desert Shield; Operation Desert Storm; Operation Northern Watch; Operation Southern Focus; Operation Southern Watch
Iraq War: A Military History, The (Murray), 153–54
Israel, 135

Jabara. James ("Jabby"), 118
Johnson, David E., 64
Johnson, Lyndon B., 123–25, 195
Jomini, Antoine-Henri, 10
Jones, Byron Quinby, 58
Journal of Military History, 200
Jusserand, Jean Jules, 38

Keegan, John, 164–65
Kelk, Jon, 150
Kenly, William, 42
Kennedy, John F., 110
Kennett, Lee, 8, 19, 46
Kenney, George C.: attendance at ACTS, 60; contributions to tactical aviation, 8, 14, 49; leadership in the South Pacific, 70, 91–92, 202; recognized for war service, 102
Knight, Clayton, 46
Korean conflict, 17–19, 90, 106, 114–20, 129, 134, 210nn37–39
Kosovo, 22, 165, 168, 172, 189, 192
Kranz, Eugene F. ("Gene"), 117
Kuhn, Thomas, 2
Kuter, Laurence S.: Air Mail scandal of 1934, 59; attendance at ACTS, 9, 12, 60, 208n13; command of MATS, 104; favoring independent air arm, 111; leadership role in USAF, 102; planning air war in WWII, 63–64,

70; preserving USAAF history, 76–78; quotation from, 100; reflection on bombing of Japan, 96; serving in North Africa, 85
Kuwait, 152–54, 156, 159, 160, 176, 180. *See also* Desert Shield; Desert Storm

Lafayette Escadrille (aka Lafayette Flying Corps), 6, 8, 27, 34–41, 43, 202, 211n23, 212nn37–38
Lafayette Escadrille Memorial Cemetery (France), 212n33
Lahm, Frank, 28–29
Lambeth, Benjamin, 178, 182
Lampert Committee, 52, 213n6
Landdeck, Katherine Sharp, 74, 215n9
Langley Air Force Base, Virginia (formerly Langley Field), 6, 60–61, 63, 154, 211n8
Leaf, Daniel P., 179
Ledwidge, Frank, 4
LeMay, Curtis: bomber command positions, 16–17, 95–96; capability as leader, 102; improved flying tactics, 85; leadership of SAC, 103, 107; military manned space flight, 195; strategic bombing strategy, 13, 19, 68, 114–15, 123
Lewis, W. David, 40
Libya, 134, 137, 139–40, 148, 201. *See* Operation El Dorado Canyon
Liggett, Hunter, 42
Limits of Airpower (Clodfelter), 21
Lindbergh, Charles, 59
Loh, John M., 156
Love, Nancy, 74–75, 202
Lufbery, Raoul, 36–38, 39–40, 212n38
Lukasik, Sebastian, 200
Luke, Frank, 46

MacArthur, Douglas, 14, 53, 56, 58, 71–72, 90–92
MacArthur's Airman: General George C. Kenney (Griffith), 14
Mahan, Alfred Thayer, 10, 11
Makers of Modern Strategy (Seversky), 11
Makers of the United States Air Force (ed. Frisbee), 52

Mannock, Edward, 31
Marshall, George C., 63, 70
Mauer, Mauer, 7, 212n43
Maxwell Air Force Base, Alabama (formerly Maxwell Field), 6, 8, 29, 59–63, 112
McCarthy, James P., 4
McConnell, James Rogers, 36, 39
McConnell, Joseph, 118
McCudden, James, 32, 211n14
McDaniel, Phil, 158–59
McDivitt, James, 194
McNamara, Robert, 124, 195
McNarney, Joseph, 8
Mediterranean Air War, The (Ehlers), 81
Meilinger, Phillip, 1, 52
Meissner, James, 7
Mets, David R., 29, 86, 97
Mexican Expedition (1916-17, aka Punitive Expedition), 29–30, 200, 214n20
Military Airlift Command (MAC): challenges of global role, 134–37; logistics of Desert Shield, 154–55; role in Desert Storm, 161–62; UASF command structure, 209n33
Military Air Transport Service (MATS), 103–5. *See also* Berlin Airlift; "flying the Hump"
Military Assistance Command-Vietnam (MACV), 120–21
Milling, Thomas DeWitt, 51
Milosevic, Slobodan, 165, 168
Mitchell, Caroline Stoddard, 55
Mitchell, Elizabeth Miller, 56
Mitchell, William Lendrum ("Billy"): airpower doctrine developed/defined, 6–7, 8–9, 11, 19, 50, 154, 194; airpower doctrine tested, 54–55; aviation leadership in AEF, 41–43; command at Battle of Saint-Mihiel, 35, 43–44; court martial, 8, 53, 56–57, 65; fight for air arm independence, 47–52; marital problems and divorce, 55; recognizing the maverick streak, 71, 92, 202; strategic bombing doctrine, 34, 49

Moffett, William, 202
Momyer, William, 121
Mongillo, Nicholas ("Mongo"), 150
Morris, Craig, 33
Morrow, Dwight, 213n6
Morrow Board (aka Board of Aeronautic Inquiry), 52
Morse, Edwin, 35–36
Moseley, T. Michael, 134
Mrozek, Donald, 202
Muehlbauer, Matthew W., 4
Murray, Williamson, 153–54, 161

Nalty, Bernard, 4, 122
National Aeronautics and Space Administration (NASA), 189, 194–96
National Cathedral (aka Washington National Cathedral), 212n33
National Defense Authorization Act, 193, 199
National Geospatial-Intelligence Agency, 198
National Museum of the United States Air Force, 126
National Reconnaissance Office (NRO), 190
National Security Act of 1947, 6, 102
National Security Council Document 162-2 (Eisenhower "New Look" policy), 16
Naval Air Transport Service (NATS), 104
Nellis Air Force Base (Nevada), 133, 141–42, 147, 148, 152, 185
Nevada Test and Training Range, 148
"New Look" (military/national security policy), 16
Nimitz, Chester, 90
9/11 terrorist attacks (September 11), 174–77, 191, 224n1
Nixon, Richard M., 122, 125–27, 135
Noriega, Manuel, 138
Norstad, Lauris, 60, 95
North African Air Campaign, The (Rein), 77
North American Air Defense Command (NORAD): creation and role, 16–17; defensive interceptor capability, 110–11, 112–13; integrated into US Northern Command, 197; integrated with Canada, 108–9; missile warning capability, 195; tracking Santa Claus, 218n18
North Atlantic Treaty Organization (NATO), Balkan air campaigns, 22, 162–72
North Korea. *See* Korean conflict
Notes on Aeroplane Fighting in Single-Seater Scouts (McCudden), 32

Obama, Barack, 75, 192
O'Conner, William, 172
O'Donnell, Emmett ("Rosie"), 115–16
Office of Air Force History, 212n43
O'Grady, Scott, 165–66
Olds, Robert, 70
Olsen, John Andreas, 4, 115, 121
Operation Allied Force (1999, bombing of Yugoslavia, aka Noble Anvil), 22, 107, 118, 164–65, 168–73, 187
Operation Anaconda (2002, Battle of Shahi Kot), 176–77
Operation Argument (1944, achieving air superiority over Germany), 86–88
Operation Avalanche (1943, invasion of Italy), 80–81
Operation Barrel Roll (1964, Vietnam War), 125
Operation Blue Spoon (renamed Just Cause), 138
Operation Cedar Falls (1967, Vietnam War), 121
Operation Cobra (1944, break-out from Normandy), 88–89
Operation Deliberate Force (1995, intervention in Republika Srpska), 22, 164–65, 167–68
Operation Deny Flight (1993, intervention in Bosnia and Herzegovina), 22, 164–66
Operation Desert Fox (1998, bombing of Iraq), 162
Operation Desert Shield (1990, Gulf War), 135, 154–55, 177

Operation Desert Storm (1991, Gulf War), 14, 21–22, 107, 118, 135, 148–56, 159–63, 167, 172–73, 178–79, 182–83, 185, 201–2, 222n4
Operation Eagle Claw, 1980 (Iran hostage rescue attempt), 148
Operation El Dorado Canyon (1986, bombing of Libya), 133–34, 137, 139–40, 148, 221n17
Operation Enduring Freedom (2001, war in Afghanistan), 107, 174–77, 184, 191, 201, 224n1, 225n1, 225n3
Operation Flax (1943, Tunisian campaign), 78
Operation Infinite Justice (renamed Enduring Freedom), 224n1
Operation Inherent Resolve (2014, war against the Islamic State), 192, 198, 200
Operation Inherent Resolve (war against the Islamic State), 175
Operation Instant Thunder, 1990 (planning the Iraqi air war), 156–58
Operation Iraqi Freedom (2003, Iraq War), 107, 153, 162, 174, 177–85, 191–92
Operation Junction City (1967, Vietnam War), 121
Operation Just Cause (1989, invasion of Panama), 133–34, 137–39, 148
Operation Linebacker I and II (1972, Vietnam War), 123, 125–27
Operation Matterhorn (1944, bombing of Japan), 93
Operation Meetinghouse (1945, firebombing Japanese cities), 95–96
Operation Nickel Grass (1973, support of Israel in Yom Kippur War), 135, 177
Operation Northern Watch (1997, enforcing Iraqi no-fly zone), 162–64, 177
Operation Overlord, (1943, Normandy landings), 86
Operation Pointblank (1943, bombing German factories), 85–86
Operation Rolling Thunder (1965, Vietnam War), 123–25, 127

Operation Saturate (1952, Korean conflict), 116
Operation Southern Focus (2002, attacking Iraqi air defenses), 178–79
Operation Southern Watch (1992, monitoring Iraqi air space), 162–64, 177
Operation Starvation (1945, mining Japanese harbors), 95, 96
Operation Steel Tiger (1965, Vietnam War), 125
Operation Strangle I and II (1951, Korean conflict), 116
Operation Tidal Wave (1943, raid on Ploesti), 85–86
Operation Torch (1942, invasion of North Africa), 77
Operation Urgent Fury (1983, invasion of Grenada), 133–34, 137–39, 148
Operation Vittles (1948, Berlin Airlift), 103–5
Over Lord: General Pete Quesada (Hughes), 10
Owen, Robert C., 105, 134

Pacific Air Forces (PACAF), 120
Panama. *See* Operation Just Cause
Parker, James E., 10
Partridge, Earle, 108
Patrick, Mason, 42, 44–47, 51–52, 55, 212n43
Pearl Harbor (Hawaii), 70–72
Pence, Michael R. ("Mike"), 193
Pershing, John J., 29–30, 41, 44, 212n43
Powell, Colin, 139
Powers, Gary, 195
Pribbenow, Merle L., 127
Prince, Norman, 35–36, 38–39, 212n33
Punitive Expedition. *See* Mexican Expedition
pursuit aviation: air-to-air combat, emergence of, 30–33; competition with bombardment doctrine, 7–10, 42, 46, 48–49; development of aircraft for, 62–63, 105–6; term redesignated as "fighters," 106. *See also* air superiority; aircraft (fighters)

Quesada, Elwood, 59–61
Quesada, Pete, 10, 13, 49, 102
"Question Mark" flight, 63
"Question Mark" flight (experimental aerial refueling), 54, 65, 135

Rainbow Plans (two-ocean war contingency plans), 63
Raymond, John, 193
Reagan, Ronald, 137
Reconsidering the American Way of War (Echevarria), 2, 3–4
Reflections on Air Force Independence (Wolk), 52
Rein, Chris, 77, 78
remotely piloted aircraft (RPA). *See* unmanned aerial vehicles/systems
Republika Srpska, 22, 164–65, 167–68
Richthofen, Manfred von ("Red Baron"), 31–32, 34
Rickenbacker, Edward Vernon ("Eddie"), 25, 31, 40, 42–43, 46, 146, 212n38
Rickman, Sarah Byrn, 74, 215n9
Rise of American Air Power, The (Sherry), 15
Risner, James Robinson ("Robbie"), 118, 203
Ritchie, Steve, 128
Robinson, Lori, 202–3
Rockwell, Kiffin, 36–39
Rodriguez, Cesar, 169
Rogers, Mike, 198
Roosevelt, Franklin D., 58, 63, 76, 92
Rostow, Walt, 124
Rumsfeld, Donald, 175, 224n1
Rusk, Dean, 124

Salter, James, 117
Santa Claus, tracking by NORAD, 218n18
Schlight, John, 121
Schriever, Bernard, 196
Schriever Air Force Base (Colo.), 198
Schwartz, Norman, 134
Schwarzkopf, Norman, 152, 156, 158, 163
Scott, Hugh, 29

Selfridge, Thomas E., 28
September 11 terrorist attacks, 174–77, 191, 224n1
Seversky P-35 (fighter plane), 63, 73
Sherry, Michael, 15, 82, 89
Short, Michael, 168
Shower, Michael K., 169
Skulking Way of War, The (Malone), 4
Slayton, Donald ("Deke"), 194
Slessor, J. C., 51
Snook, Scott A., 164
Solona, Javier, 168
Sorenson, Edgar P., 76
Spaatz, Carl Andrew ("Tooey"): circumnavigation of the globe, 54; court martial of Mitchell, 57; graduation from ACTS, 60; leadership that shaped the Air Corps, 46, 65; named USAF commander, 101–3; reflecting on bombing of Japan, 96–96; serving in Mexican Expedition, 29–30; serving in WWI, 45; serving in WWII, 70, 73, 78–79, 89, 210n43; spelling of last name, 208n13, 211n8
Stafford, Thomas, 194
Stealth Fighter (O'Conner), 172
stealth technology, development of, 147–48
Steiner, Carl, 138
Stephens, Alan, 8–9, 119
Stewart, James ("Jimmy"), 107
Stewart, James T., 119
Story of the Lafayette Escadrille, The (Thenault), vii, 36
Strategic Air Command (film), 107
Strategic Air Command (SAC): creation and role, 15–17, 107–9; nuclear weapons delivery, 102–3, 107, 110, 119–20, 196; reorganized as Global Strike Command, 226n14; role in Korean Conflict, 115; role in Vietnam, 120–22; strategic bombing doctrine, 19–20, 112. *See also* Tactical Air Command; United States Air Force
strategic bombardment: Combined Bomber Offensive (CBO) and, 77–85,

89; creation and role of AWPD-1, 63–65, 67–70, 91, 104; doctrinal development between wars, 48–51; doctrinal development in WWI, 33–34, 42; doctrine problems in Korea and Vietnam, 19–22, 114–17, 119, 128; dominance in air doctrine, 7–19, 60–61, 133; employment in North Africa, 67–70; employment in European Theater, 81–90; employment in Pacific, 90–99; recognizing contributions in WWII, 100; shift to nuclear weapons delivery, 100–103, 112; shifting delivery to fighter and attack aircraft, 154; technology and implications for, 62, 201–2
Stratemeyer, George, 17, 60, 93, 102
Structure of Scientific Revolutions (Kuhn), 2
Sullivan Glen, 126
Sun Tzu, 10
Suter, Moody, 203
Sutherland, Richard, 71–72
Syria. *See* Operation Inherent Resolve

Tactical Air Command (TAC): creation and role, 15–16, 102–3, 134–37, 209n33; dominance of the SAC over, 19–21, 119–20; fighter capabilities, 109–10; lessons learned in Korea, 17–18; lessons learned in Vietnam, 21, 128; training and employment of aircraft, 140–47
Tactical Analysis Bulletin, 144–46
Taft, William Howard, 28–29
Talbott, Harold, 111
Tate, Steve, 150
Taylor, Kenneth M., 71
Thaw, William, 35–40
"The Air Force and the Future of the Strategic Bomber" (Werrell), 19
"The Deliberate Force Air Campaign" (Campbell), 167
Thenault, Georges, vii, 36–38, 40
Thornhill, Paula, 5
Thunderbirds (Aerial Demonstration Team), 106

Thurman, Maxwell R., 139
Tibbets, Paul, 14, 81–82
Tilford, Earl H., Jr., 124
Time (magazine), 102
"To B(FM) or not to B(FM): That is the Question" (High), 146
Tonopah Test Range (Nevada), 148
Toomey, Dave, 180–81
Trenchard, Hugh, 203
Trump, Donald J., 193, 197, 199
Tsiolkovsky, Konstantin, 194
Tunner, William H., 60, 93, 104, 202
Tuskegee Airmen, 203
Twelve O'Clock High (film), 82

Ulbrich, David J., 4
United Nations, support of military operations, 17, 115, 154, 165, 167, 168
United Nations Resolution 660, 154
United States Air Force (USAF): achieving independence as air arm, 1, 4, 8, 11, 63–65, 100–101; bombardment as dominant mission, 15–16, 68, 99, 103; creation of, 102; dealing with Cold War challenges, 59, 103–9, 114, 136; developing ICBM force, 103; developing officer cadre, 59–60, 111–12; developing doctrinal autonomy, 79–80; developing new generations of aircraft, 105–11; developing technology and science, 101–2; Fighter Weapons School, 140–46, 148; fixing doctrinal failures following Vietnam, 21; preparing for future challenges, 185, 187–90, 203; program development, ICBMs, 196–97; program development, satellites and manned missions, 194–96; recognizing mavericks and innovators, 202–3; space responsibilities turned over to USSF, 192–94. *See also* Strategic Air Command
United States Air Force Academy (USAFA), 3, 111, 113
United States Army Air Arm, April 1861–April 1917 (Hennessy), 34

United States Army Air Force (USAAF): bombing doctrine, 9, 13; creation and role, 70; forming officer cadre, 70; gaining air superiority over Germany, 85–88; importance of tactical aviation, 13–14; launching European campaign, 81–85; passage of Army Air Corps Act of 1926, 51–52; preserving history of, 76–77; supporting Operation Avalanche, 80–81; supporting Operation Cobra, 88–89; supporting Operation Torch, 77–80; supporting CBI Theater ("flying the Hump"), 92–93; taking war to Japan, 93–97; transitioning to post-war period, 100–103; women, role in WWII aviation, 75
United States in the Air, The (Patrick), 52
United States Space Force (USSF): creation and role, 5, 22–23, 188–90, 193–94; establishment as separate force, 197–99; NORAD attachment to, 197. *See also* Air Force Space Command
United States Strategic Bombing Surveys, The (USSBS), 13–15, 90, 96–97, 209n26
University of Oklahoma Press, 3
unmanned aerial vehicles/systems (UAVs/UAS): changing way of making war, 22, 185, 187–88; early development and use, 190; use in Allied Force/Noble Anvil, 171; use in Enduring Freedom, 175, 191; use in Iraqi operations, 191–92; use in Balkans campaigns, 167, 191–92
Unseen War, The (Lambeth), 178
U.S. Air Service in World War I, The (Mauer), 7, 42, 212n43
US Army Air Corps Act of 1926, 51–53
US Army Signal Corps, 6, 28, 41
US Central Intelligence Agency (CIA), 175, 190
USS *Langley*, 72

Vandenberg, Hoyt S., 9, 10, 60, 102, 115, 119, 192–93

Verne, Jules, 33, 34
Victory in Vietnam: The Official History (Pribbenow), 127
Victory through Air Power (1943, film), 12, 19
Victory through Air Power (de Seversky), 5–6, 11–12
"Victory through Hot Air Power" (Brown), 11
Vietnam War, 1, 19–21, 89–90, 107, 109, 114–17, 120–29, 190
Villa, Francisco ("Pancho"), 29–30
Von Kármán, Theodore, 101

Walker, Kenneth, 12, 63, 70, 91–92
Warden, John, 21, 133, 153, 156–57, 182–83, 202, 208n15
WarGames (movie), 218n18
War in South Vietnam (Schlight), 122
War in the Air, The (Wells), 33
War on Terror. *See* 9/11 terrorist attacks
"way of war," American airpower and the: defining concept, 1–7, 200–203; 1907–1941, WWI and birth of, 7–10, 26, 33, 35, 48, 64; 1942–1975, strategic bombing, dominance of, 10–20, 68, 74, 92–94, 98, 107, 112, 118, 129; 1975–2019, tactical air force, ascendancy of, 20–22, 132, 133, 137, 141, 164, 167, 173, 174, 185; 2019–to present, UAVs and future of, 22–23, 188, 189–90, 193, 199
Ways of War: American Military History (Muehlbauer and Ulbrich), 4
Weigley, Russel, 3
Welch, George, 71
Wells, H. G., 33, 34
Wells, Mark, 98–99
Werrell, Kenneth, 19
Westmoreland, William, 122
Westover, Oscar, 51, 53
Weyland, Otto Paul, 16
Wheeler, Earle, 124
Wheeler Army Air Field (Hawaii), 70–71
White, Ed, 194
Whitman Air Force Base (Missouri), 172
Wilson, Heather, 193, 199
Wilson, Woodrow, 212n33

Winged Defense (Mitchell), 6, 48, 57
Winged Shield, Winged Sword (Nalty), 4
women: aviation, wartime role in, 5, 65, 99, 202–3; becoming fighter pilots, 131, 146, 201; casualties and death from bombing, 13, 62, 96; flying combat missions, 76, 202–3; pilots, wartime training as, 21, 74–76, 215nn9–10; Space Force, role in, 197
Women Airforce Service Pilots (WASP), 74–76, 202, 215n9
Women's Auxiliary Ferrying Squadron (WAFS), 74–75
Women's Flying Training Detachment (WFTD), 74–75
Worthington, Walter, 139
Wright, Orville, 28–29, 59
Wright, Robert ("Wilbur"), 165–66, 169
Wright, Wilbur, 28–29
Wright Brothers Flying School, 6

Yom Kippur War, 135
Yugoslavia, 22, 165–69, 171

Zelko, Dale, 169–70
Zhang, Xiaoming, 118

www.ingramcontent.com/pod-product-compliance
Lightning Source LLC
Chambersburg PA
CBHW021343230426
43666CB00006B/387